ANATOMY &
100 STRETCHING EXERCISES FOR
RUNNERS

BARRON'S

ANATOMY &
100 STRETCHING EXERCISES FOR
RUNNERS

All inquiries should be addressed to:
Barron's Educational Series, Inc.
250 Wireless Boulevard
Hauppauge, NY 11788
www.barronseduc.com

ISBN: 978-1-4380-0719-9

Library of Congress Control No.: 2014955096

Editorial Director: Maria Fernanda Canal
Text: Guillermo Seijas Albir
Proofreading: Ana Lorenzo
Typographical Correction: Roser Pérez
Graphic Design: Toni Inglès
Illustrations: Myriam Ferrón
Photographs: Nos i Soto
Layout: Estudi Toni Inglès

Paidotribo is grateful for the collaboration
on this book with: Anna Baeza Dalmau,
Sergi López Borrego, Noemi Morales
Lorenzo, and Jordi Pradell Pascual

Printed in Spain
9 8 7 6 5 4 3 2

First edition for the United States, its
territories and dependencies, and Canada
published in 2015 by Barron's Educational
Series, Inc.

English-language translation © copyright
2015 by Barron's Educational Series, Inc.
English translation by Eric A. Bye, M.A.

Original Spanish title: *Anatomía & 100
Estiramientos Esenciales para Running*
© Copyright 2015 Editorial
Paidotribo—World Rights
Published by Editorial Paidotribo,
Badalona, Spain

There are few things that provide you with as much pleasure and well-being as running. The pride in having met a tough challenge, the satisfaction that comes from finishing a workout and knowing that you have given it your all, going beyond your limits, and the continual struggle to do better, are very satisfying. There is no sensation or activity that makes you feel more complete. In this sense, every sport is gratifying, but few of them give back as much in exchange for so little as running does. You can put on your running gear any time and go out for a run, without lengthy preliminaries, without complex equipment, without special facilities or barriers between you and the asphalt, dirt road, or open countryside.

Its simplicity, its versatility, and its appropriateness for every athlete and level make running one of the best means for personal achievement. Whether in the country, in the city, or at the beach, it matters not: all you have to do to be a runner is to take the first stride. Maybe you can last only five minutes the first time, but progress and improvement will not be long in coming, and soon you will be able to keep up a running pace for thirty or forty minutes, take part in informal races, or who knows what—you can set your own limits.

But don't fool yourself: every athletic endeavor involves some risk, and running is no exception. Over time, you may experience joint problems or muscle strain, or simply discover that you have reached your maximum potential, and this may discourage you or cause you to quit the sport. So, it is important to have the necessary knowledge before starting, and to use proper methods and techniques to reduce these risks.

In these pages, you will find: technical concepts on running that will help you improve your style and performance by developing an optimum economy for running; an analysis of the basic biomechanics of running and of the muscle requirements at every moment; and a complete selection of the best stretching exercises for the physical demands involved, with special emphasis on the muscle groups that are most prone to strain and fatigue. You will learn the best way to stretch at every moment, how to do the various stretches without incurring unnecessary risk, and which exercises may serve you best in each of the most popular running disciplines.

Finally, in order to avoid delaying for a moment the start of your running, we include a selection of pre-designed stretching workouts based on the time you have available for doing them.

All this information is designed to improve your athletic experience and your performance, and, especially, to help you enjoy a long and complete athletic career as a runner.

Contents

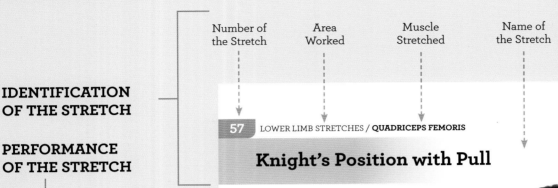

Number of the Stretch

Area Worked

Muscle Stretched

Name of the Stretch

IDENTIFICATION OF THE STRETCH

PERFORMANCE OF THE STRETCH

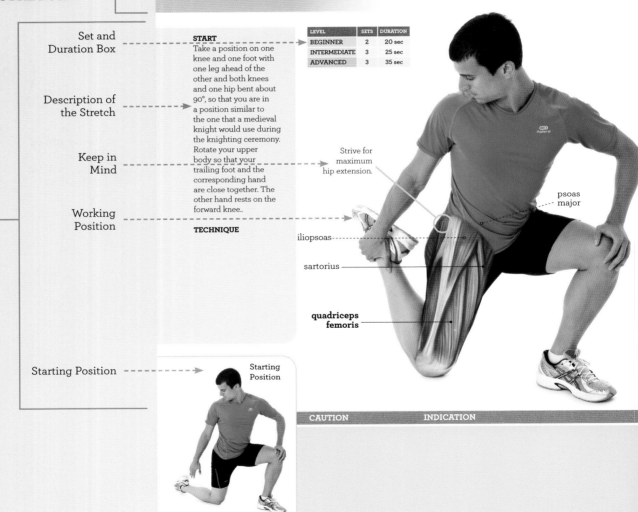

57 LOWER LIMB STRETCHES / **QUADRICEPS FEMORIS**

Knight's Position with Pull

Set and Duration Box

Description of the Stretch

Keep in Mind

Working Position

Starting Position

START

Take a position on one knee and one foot with one leg ahead of the other and both knees and one hip bent about 90°, so that you are in a position similar to the one that a medieval knight would use during the knighting ceremony. Rotate your upper body so that your trailing foot and the corresponding hand are close together. The other hand rests on the forward knee..

TECHNIQUE

LEVEL	SETS	DURATION
BEGINNER	2	20 sec
INTERMEDIATE	3	25 sec
ADVANCED	3	35 sec

Strive for maximum hip extension.

psoas major

iliopsoas

sartorius

quadriceps femoris

Starting Position

CAUTION   INDICATION

Page Number and Chapter

98/ Static Stretches After Running

One stretch per page, like an index card

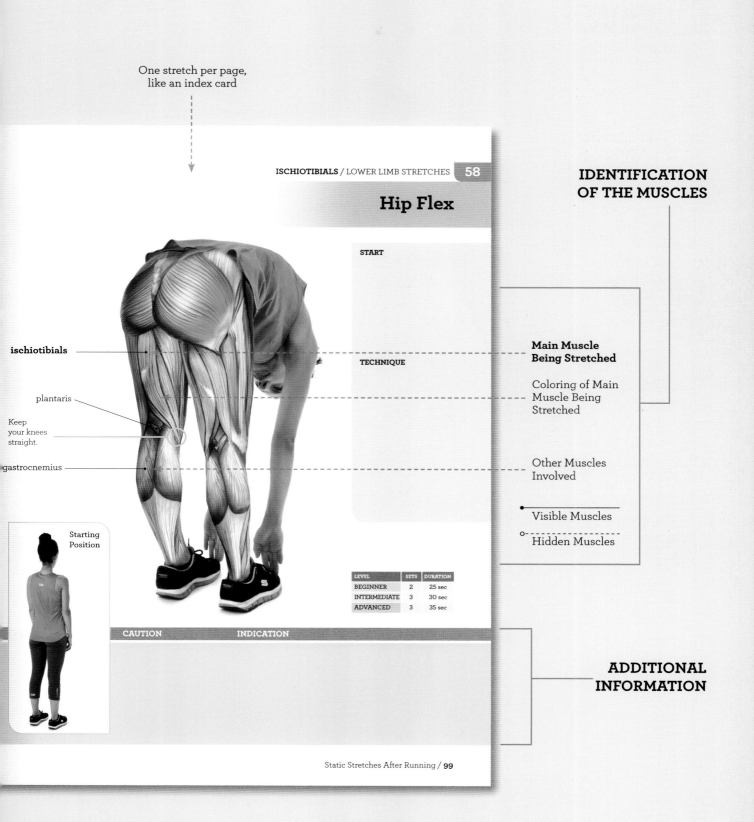

Hip Flex

START

ischiotibials

plantaris

Keep your knees straight.

gastrocnemius

Starting Position

TECHNIQUE

IDENTIFICATION OF THE MUSCLES

Main Muscle Being Stretched

Coloring of Main Muscle Being Stretched

Other Muscles Involved

● Visible Muscles

○ Hidden Muscles

LEVEL	SETS	DURATION
BEGINNER	2	25 sec
INTERMEDIATE	3	30 sec
ADVANCED	3	35 sec

CAUTION   INDICATION

ADDITIONAL INFORMATION

Static Stretches After Running / **99**

Anatomical Atlas
Location of the Muscles

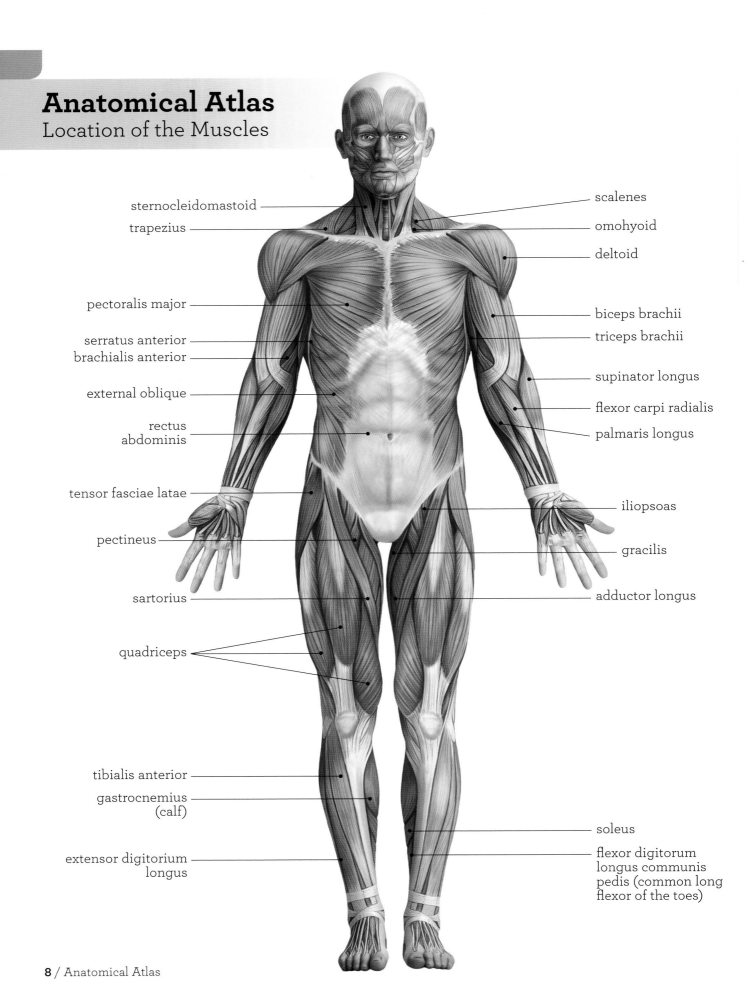

sternocleidomastoid

trapezius

pectoralis major

serratus anterior

brachialis anterior

external oblique

rectus
abdominis

tensor fasciae latae

pectineus

sartorius

quadriceps

tibialis anterior

gastrocnemius
(calf)

extensor digitorium
longus

scalenes

omohyoid

deltoid

biceps brachii

triceps brachii

supinator longus

flexor carpi radialis

palmaris longus

iliopsoas

gracilis

adductor longus

soleus

flexor digitorum
longus communis
pedis (common long
flexor of the toes)

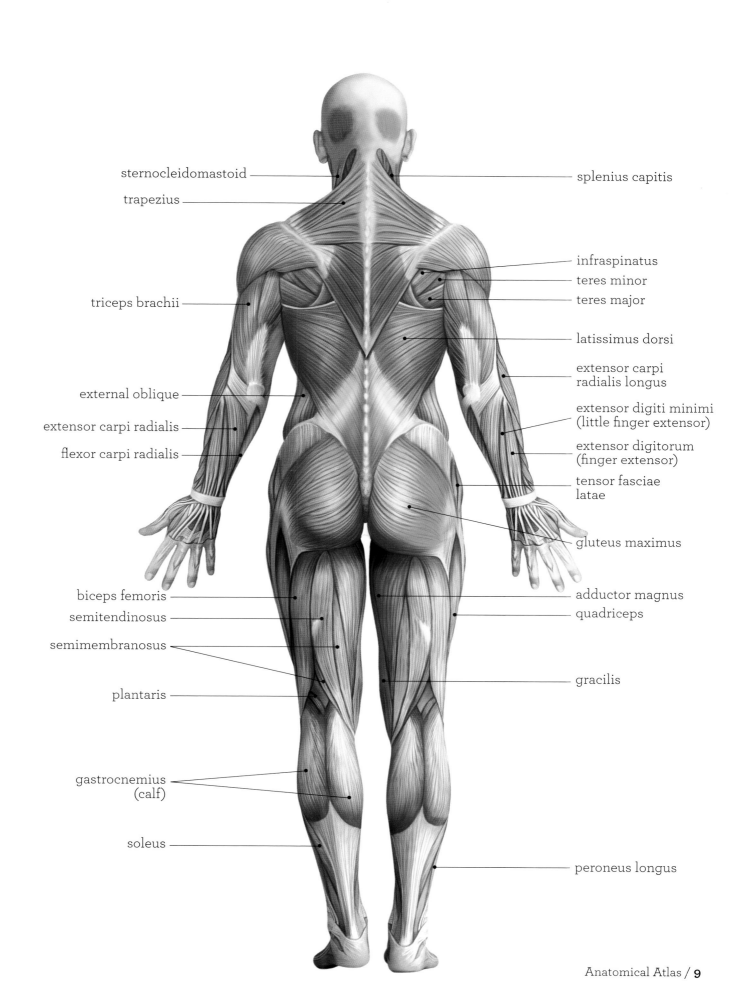

sternocleidomastoid

trapezius

triceps brachii

external oblique

extensor carpi radialis

flexor carpi radialis

biceps femoris

semitendinosus

semimembranosus

plantaris

gastrocnemius
(calf)

soleus

splenius capitis

infraspinatus

teres minor

teres major

latissimus dorsi

extensor carpi
radialis longus

extensor digiti minimi
(little finger extensor)

extensor digitorum
(finger extensor)

tensor fasciae
latae

gluteus maximus

adductor magnus

quadriceps

gracilis

peroneus longus

Planes of Movement

Before starting, it is appropriate to explain a series of terms that refer to body movements and that will appear consistently throughout the book. If you do not know the basic nomenclature of the movements, it will be difficult to understand the detailed descriptions of the exercises. Some of these terms—such as *flexion* and *extension*—are commonly used, but others—such as *inversion, eversion, adduction,* and *supination*—are used less often, so it can be very helpful to clarify their meaning.

The first thing we need to know is that body movements occur in three different planes: the frontal, the sagittal, and the transverse. Each of these planes involves a certain series of movements, as we will see. To understand them, we can begin with the basic anatomical position, which is shown in the illustration.

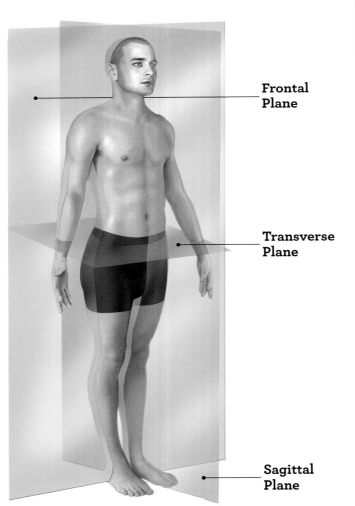

Frontal Plane

Transverse Plane

Sagittal Plane

ABDUCTION

ADDUCTION

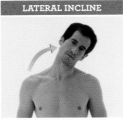
LATERAL INCLINE

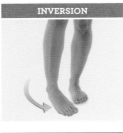
INVERSION

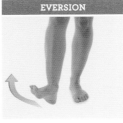
EVERSION

FRONTAL PLANE

This plane divides the body into ventral and dorsal parts—in other words, anterior and posterior. The chest and stomach are in the ventral part, and the back of the neck, the back, and the buttocks are in the dorsal part. The movements of the frontal plane are the following:

Abduction: This is the movement in which we move a limb away from the central axis of the body. It is easily seen from the front or the rear of the individual, because the change in silhouette is obvious from that vantage point. In forming the shape of a cross with your arms, you perform abduction of the shoulders.

Adduction: This is a movement by which we bring a limb toward the central axis of the body—in other words, the movement that is opposite abduction. If your arms form a cross and you lower them so they are next to your body, you perform an adduction of the shoulders.

Lateral Incline: In this movement, the head, the neck, or the upper body is tilted to one side. If you fall asleep while sitting up, generally your head and neck eventually tip to one side through lateral inclination.

Inversion: Although this movement does not involve the frontal plane exclusively, this is the plane where it is most common. Inversion of the foot results when the tip and the sole are turned inward at the same time plantar flexion is performed.

Eversion: This is the movement in which the tip and the sole of the foot are turned outward at the same time that the foot experiences dorsal flexion.

FLEXION

EXTENSION

ANTEPULSION

RETROPULSION

DORSAL FLEXION

PLANTAR FLEXION

SAGITTAL PLANE

This plane divides the body into two halves, the right and the left. The movements in this plane are best perceived from one side of the individual, in a profile view. In this plane the following movements stand out:

Flexion: This is the movement in which we advance a part of the body with respect to the central axis. For example, if you bend your elbow, you move the forearm ahead with respect to the central axis. There are exceptions to this definition, such as flexion of the knee and plantar flexion of the ankle.

Extension: A movement in which we shift a part of the body rearward with respect to the central axis, or in which we align a body part with the axis. For example, if you look at the sky while you are standing, you have to perform an extension of the cervical column. Once again, the knee is an exception.

Antepulsion: This is the equivalent of flexion, but it applies solely to shoulder movement.

Retropulsion: This is equivalent to extension, but it applies solely to shoulder movement.

Dorsal Flexion: A flexing movement applicable solely to the ankle joint.

Plantar Flexion: This term designates the movement of the ankle that is equivalent to extension.

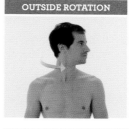

OUTSIDE ROTATION

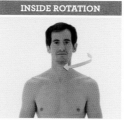

INSIDE ROTATION

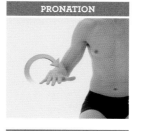

PRONATION

SUPINATION

TRANSVERSE PLANE

This plane divides the body into upper and lower parts. Movements in this plane are easily perceived from any angle, but slightly better from the rear or from beneath the person. They include the following:

Outside Rotation: This is the movement in which we turn a part of the body outward and along its own axis. If you are seated at a table and a fellow diner next to you speaks to you, you perform an outside rotation of the neck to look at the person.

Inside Rotation: This is the movement opposite the preceding one, because it involves turning toward the inside and along the axis of a body part. As you finish the conversation with a fellow diner seated next to you, you perform an inside rotation of the neck to return your gaze toward the front.

Pronation: A rotational movement of the forearm, by which we place the back of the hand upward and the palm downward. When you use a knife or a fork to manipulate the food on a plate, your hands are in pronation.

Supination: This is the movement opposite the preceding one. It involves the rotation of the forearm, in which we place the palms facing upward. For example, if someone gives you a handful of seeds, you place your hands palms up, in supination, like a bowl, so you don't drop them.

The Origins of Running

Running is currently one of the most popular sports. It is a phenomenon that motivates hundreds of thousands of people all over the world and that involves dozens of disciplines and modalities. The variety of events depends not only on the distances covered, but also on variations in equipment, types of surfaces, the presence or absence of obstacles, gradients, and even combinations with other disciplines.

There's nothing new about the action of running. Our ancestors, the first ones to move about on two feet, used walking and running for survival. The first hominids used running for avoiding their predators as well as for hunting their prey, and that included covering long distances in search of areas where food and water were more abundant.

At first glance, it may seem that humans are not the animal best adapted to running, because the fastest humans in the world, such as the sprinter Usain Bolt, can barely reach 24–27 mph (40–45 km/h), and people who run on a lower level rarely exceed 21 mph (35 km/h). This is not a particularly high speed, and it may even be frustrating if, for example, we consider that a cat can reach about 29 mph (48 km/h) and a rhinoceros 24 mph (40 km/h)—and this without mentioning other species, such as the lion, which can reach 48 mph (80 km/h), and the leopard, with its astonishing 68.5 mph (114 km/h). The fact is that even animals that barely reach a tenth of our weight and size can run faster than we can. So, if we consider that most of our ancestors' potential predators and prey were speedier than we are, it seems logical that, if one day we meet a predator face to face, it would be better to climb a tree than to take off running.

The leopard is one of the swiftest animals, and it can reach a maximum speed of 68.5 mph (114 km/h).

The genetic background
and environmental
conditions of Kenyans and
Ethiopians make them
exceptional runners.

As we have seen, the advantage that our ancestors had in fleeing was not speed, but rather their bipedal posture, which allowed them to scan the horizon and detect danger long before it came so close that they had to run away.

Humans are natural-born runners, but not sprinters. The first hominids caught their prey not because of their speed, but because of their endurance and, of course, their intelligence. Humans are in fact capable of covering very long distances at a trot, and this differentiates them from most land animals.

Nowadays, there are groups of people who have made running a kind of lifestyle. The Kalenjin ethnic group in Kenya, who live in the Rift Valley, are one example of this. Throughout their history, they have covered great distances on foot in order to do the simplest daily tasks, such as going to fetch water or to carry a message to a neighboring town. The need for running, along with life at high altitude, has made them ideally suited to running, to the point that they are an inexhaustible source of champions in international long-distance races. Kenya and Ethiopia are the birthplaces of great marathon runners, such as Haile Gebrselassie and Patrick Makau, who are living proof that humans are long-distance runners.

On the other side of the world, in the western Sierra Madre Mountains of Mexico, there is another group of people who are particularly adept at running, the Tarahumara (or Raramuris), whose environmental circumstances are similar to the Kalenjin runners. For starters, they also live in poorly connected areas with few resources, so they have to cover long distances on foot to get the basic necessities for daily life. In addition, because their homeland is mountainous, many inhabitants live at high altitude—for example, in Guachoci, which is located at nearly 8,000 feet (2,400 m) above sea level. This has contributed to the inhabitants' adaptation to an environment with relatively little oxygen. This has enabled the Tarahumara to take in and utilize oxygen more efficiently than the rest of us, which makes them superb long-distance runners.

*The Tarahumara, or "light feet,"
run long distances wearing
simple sandals.*

All of this is augmented by a cultural trait in the case of the Tarahumaras, although it is not unconnected to the hard life that the environment imposes on them: one of the most important collective activities that they engage in is a ball game in which several individuals sometimes run up to 120 miles (200 km) as they kick a ball with their feet. The game may go on for up to two days and nights without a break for the runners. And, even though the speed is lower than the average for a marathon runner, the difficulty of the game is undeniable. There's a reason that Raramuri means "light-footed people."

So, we cannot ignore the fact that running has played a significant role in the origin and development of humans. People have used running to move to areas richer in water and food, to flee from dangers, and to hunt, managing to accomplish this last task by completely wearing out their prey.

Although there still are populations for which running is vitally important, given the difficulty of their environments, in the developed world people move about in relative ease, and there are no longer great dangers, or, at least, there are no dangers from which people can't get away by running. But this has not diminished the natural instinct for running that people show from their infancy. Running is our favorite game from the time we manage to stand, and it extends through time in a multitude of athletic disciplines. We are not referring exclusively to running competitions: we run in soccer, basketball, baseball, handball, and field hockey, and we could continue mentioning physical and athletic activities that involve running *ad infinitum*. In fact, running is part of the vast majority of athletic activities that we engage in on land, even though sometimes it involves only a few strides.

In any case, when people have not had to run in order to survive, they have run for enjoyment. As in ancient Greece and the Egypt of the pharaohs, sports and running have become forms of play, competition, and even a way to obtain social prestige.

Nowadays, running is a rising phenomenon, and, technically, it is a simple thing to do: even people who put on running shoes for the first time can keep up a running pace for a couple of minutes. It is an athletic pursuit within everyone's reach, because all you have to do is go out to the road for a run. And, if you want to use any particular equipment, there's not much of it and it's readily available.

The combination of running and leisure time has led to dozens of kinds of activities, including very traditional forms of running connected to track, such as 100-, 200-, and 400-meter sprints; 5,000- and 10,000-meter runs, the half-marathon, and the full marathon.

There are also a great many popular events with open participation and a variety of distances and difficulties, including some that take place inside the subway tunnels of certain cities.

On the other hand, ultra-long distance and cross-country events have seen an increase in the number of aficionados, despite the fact that they are most likely to involve risks due to their intensity.

Cross-country is an athletic discipline conducted over a natural course. Even though the distances are not terribly long, the hills and obstacles that may exist in the terrain make them very difficult competitions. Nature lovers prefer these to races on asphalt, such as half-marathons.

Another option for people who prefer natural landscapes and slightly cleaner air is trail running, or mountain running, which is a harsh cousin of cross-country, because the distances covered are considerably greater and the courses are rougher. They may involve trails and secondary roads, and may even require crossing streams. These events don't have an established distance, but they commonly exceed six miles (10 km), and they may be longer than 100 miles (160 km) in ultra-trail-running events.

We should also mention mountain- and sky-running, which are events conducted at an altitude of over 6,600 feet (2,000 m). They require excellent training and physical conditioning, given the lack of oxygen, even though they may be child's play to the Tarahumara. Kilian Jornet, a Catalan athlete with a long career in ultra-long-distance events, is a standout in this discipline.

This is just a brief summary of events to provide some context, and clearly there are many other events that we have not described, even though they are just as important as the ones mentioned here and even though they have a great many followers and devotees.

Finally, we should point out that, even though running is a sport for everyone, it must be done carefully and with respect for the sport and the environment. Running is fun, and it contributes to improved physical conditioning, but it is also an activity with tremendous impact on the joints, particularly the knees.

It is highly recommended to use athletic shoes with a cushioned sole to dampen the repeated impact involved in running, and you should examine the treads and use custom-made shoes if you are going to cover more than 18–24 miles (30–40 km) a week. In this regard, the professional advice is simple: shun fashion.

Don't use shoes with curved soles, unless you see elite athletes using them in official competitions, which will not happen. Don't run barefoot, unless you do it on the African savanna and you have protective calluses at least 3/8 inches (1 cm) thick over the entire soles of your feet. Don't run on tiptoes, because your Achilles tendons are not made of titanium and they will not be able to withstand stress that is likewise too much for your knees. And don't have blind faith in publications whose sole purpose is to sell you products for use in running.

The last bit of advice: if someone tells you that, in order to be more natural, you have to run the way our ancestors did, remember that our ancestors walked around naked, ate a lot just once a day, and had a life expectancy of less than 30 years. Do what is best for your health while getting the benefits of a running regimen.

Running shoes should be light and offer good cushioning in the sole to absorb the impact of every footfall.

The Biomechanics of Running

Like all athletes who wish to improve their performance, avoid injuries, and prolong their athletic careers, runners need to be familiar with the basic technical elements of running. You don't have to be a top-level competitor or a professional athlete. Anyone who intends to run sporadically, or even regularly but recreationally, needs to know about running technique and the factors that influence it.

How many aficionados of racquet sports have suffered from "tennis elbow?" If you consult the research, you will see that the percentages are surprising. This ailment cannot be avoided in all cases; but, in many cases, it could be minimized or dealt with, given adequate preparation and some basic knowledge of the biomechanics of tennis, paddleball, squash, and other such sports. In addition, it could be very efficient and productive in your intended sport to have at least a rudimentary knowledge of how your body functions when it is exercised.

This can apply to any physical activity, including running. Obviously, it's the simplest thing to put on your running shoes and go out for a run without further ado, and there's nothing wrong with that, because, in the final analysis, the most important thing is to move. If the dilemma is between

reading a book on running or going out for a run, always choose the latter, even though it's highly probable that you could do both. But, if you start running more or less regularly, no matter how modest your pace, it's a good idea to acquire some knowledge about what you are doing, and thus avoid possible injuries and wasting money on equipment you don't need.

In fact, you only need to focus on the phrase attributed to Socrates: "Knowledge will set you free." Maybe knowledge won't make us totally free, but, of course, it can save us from some injuries. And, above all, it will help us make informed choices of a running regimen that meets our needs. The great champions don't owe their success solely to their unique genetics, their persistence in training, or their modern shoes, which are heavily advertised in the media.

Ground phase.

Maybe that's the way things were 50 years ago; but, nowadays, there is a lot of science behind the sport: all high-level athletic trainers base their methods on a broad knowledge of how the mind and body function.

If this works for Maurice Greene and Paul Tergat, why should it not work for you?

Today, the information is far more readily available, and, even though most of us can't afford a star trainer, we certainly can make ourselves into our own best trainers.

This does not mean that there is a single, perfect way to run, but there are some basic parameters for efficient running to which all runners should adjust their own style. No two runners are exactly alike, and nobody has a corner on the perfect style. Just as every person is different and unique, so is every runner, and everyone needs to find the best individual way of running, within certain general guidelines.

So, the next step involves knowing what happens when we run, which muscles work at every moment, which ones can withstand greater stress, and, ultimately, which ones need to be worked intensively in stretching sessions. Armed with the information in this book, you'll be able to put together an effective stretching regimen.

Running involves different phases. There is a simple distinction between the ground and aerial phases of running.

GROUND AND AERIAL PHASES

The ground phase is when the runner has one foot in contact with the ground. This is the most interesting phase to analyze at the biomechanical level, because this is where the greatest stresses and applications of force take place—and these involve great muscle power and exertion on the part of the runner—and, in large measure, they determine the efficiency of the running style. Because this phase is especially important, we will analyze its various components in detail later on.

The aerial phase is when the runner is not in contact with the ground, but rather is "airborne." Walking is different from running in that it has no aerial phase, because there is always at least one foot in contact with the ground. This is totally natural in our usual walking style, and there is nothing striking about it; but, if you watch a walking competition attentively, you will see the hip rocking that is characteristic of walkers' technique, which allows them to move quickly while still keeping one foot in contact with the ground (otherwise, they would be penalized). The length of the aerial phase depends to a great degree on the type of event. The higher the speed, the longer the aerial phases, and the longer the flight. Long-distance events involving a slower speed necessarily entail a greatly reduced aerial phase.

Aerial phase.

During the ground phase of running, there are several moments that we will analyze in detail, because they will help to determine which muscle groups withstand greater tension, and thus are the ones that we need to focus on both in stretching sessions and in possible strengthening work, especially in the case of sprinters or others who regularly experience strain in specific areas.

LOWER BODY

Contact phase or initial support and shock absorption: This occurs at the instant when the forward foot makes contact with the ground, and marks the end of the aerial phase and the start of the ground phase. The foot should make contact with the central side area of the sole, the metatarsal area, although many runners make contact slightly farther back. This contact may vary as a function of the running regimen, moving the support toward the front of the foot, mainly during acceleration in sprinting events. To make this first contact as described, the dorsal flexors of the ankle must come into play, especially the tibialis anterior. This first contact with the central side area, or the posterolateral area of the foot, becomes a support for the entire sole by means of plantar flexion in the ankle, which is always conditioned by the activity of the **tibialis anterior**. This sequence is one of the shock-absorbing elements of this first contact. Secondly, contact is never made with the knee totally straight, because that would be very traumatic for all the elements of the joint. Because there is a slight bend in the knee, it would tend to bend completely after the initial contact with the ground. To counter this tendency, which surely would cause the runner to fall, the extensor muscles of the knee must limit the bend, and, at this instant, the **quadriceps femoris** plays a major role. Thus, this slight play in knee movement is another factor that contributes to dampening the force of the initial contact with the ground. Finally, the hip is partially flexed at the instant when contact is made, and it too would end in total flexion without the intervention of the hip extensors—the **gluteus maximus**, the **gluteus medius**, and the **ischiotibial (hamstring) muscles**. So, the hip is the third shock-absorbing element in the initial support.

Middle support phase: This occurs between initial support and takeoff, and it involves the part of the movement in which the support limb is nearly perpendicular to the ground. In this phase, the main support task occurs through the nearly isometric contraction of the **quadriceps femoris**, which prevents knee flexion, and, thus, an excessive lowering of the center of gravity. This element and proper initial contact keep the body from dipping and rising at every stride, so that it maintains a horizontal line and the energy is invested in forward movement. Also, although to a lesser extent, the hip extensor muscles keep working to increase hip extension. The main ones used in this action are the **gluteus maximus** and the **ischiotibials**. As the runner approaches the final part of the middle support phase, the plantar flexors of the ankle come into play, mainly the **gastrocnemius (calf**

The contact, or initial support, phase (A), middle support (B), and final support (C).

A

B

C

muscles) and the **soleus**. They prevent dorsal flexion of the ankle and contribute to the initial separation of the heel from the ground.

Final support phase, thrust, and takeoff: This involves the end of contact between the support leg and the ground, where the runner creates the thrust that produces the aerial phase. This requires considerable effort by the hip extension muscles, especially the **gluteus maximus** and the **ischiotibials**, to create a quick and maximum extension. In addition, the tension in the ischiotibials allows controlled extension and stability of the knee, which gives it the rigidity it needs to function as an efficient lever. Finally, the action of the **gastrocnemius** and **soleus** muscles facilitates the powerful plantar extension of the ankle, contributing significantly to the forward thrust of the body. This thrust sequence is common to all events; but, clearly it is crucial in the sprinting events, in which the speed of movement and the power of the muscle groups are essential for creating the best results.

The swing phase: The purpose of this phase is the recovery of the trailing leg, or, what amounts to the same thing, to advancing the leg used for the thrust to the forward position prior to the initial contact, at which point the cycle begins anew. In contrast to what happens in the phases affecting the lower body, the swing occurs during both the aerial and the ground phases. When you begin to recover the thrust leg, you are still airborne, with no support. The first contact

usually occurs just before the thighs come into alignment. Subsequently, the swinging leg continues moving forward until it is ahead of the upper body. During this process, the hip passes from maximum extension to a considerable degree of flex, which can reach a 90°-angle with respect to the upper body in some sprinters. Then the angle diminishes progressively. This movement occurs due to the main action of the **psoas major**, even though, as we have pointed out, no great application of force is required. The knee performs a similar sequence, passing from total extension at the end of the thrust phase to a very pronounced flex in the middle section of the swing produced by the action of the **ischiotibial** muscles, and then it straightens again before making contact with the ground. This last action is accomplished by the **quadriceps femoris**. As for the ankle, it goes from maximum extension as a result of the thrusting action to a neutral position.

Finally, but just as significantly, we need to keep in mind that there are other muscles at work in the lower body, even though they are not the main ones involved in producing the movements of running.

These muscles perform stabilizing functions, mainly in the hip and calf, so we need to include them in our training programs. The actions of the adductors, the pectineus, the gracilis, the tensor fasciae latae, and the gluteus medius and minimus are important in stabilizing the hips. The peroneus and the tibialis posterior are crucial in stabilizing the ankle.

These three photos correspond to the swing phase.

THE UPPER BODY

Even though the upper body is not the main driver in running, it certainly plays a basic role in both stability and forward thrust. Can you imagine what it would be like to run with your elbows straight and your arms in constant contact with your body? Try it—when there's nobody else around—and you will see that, even though it is possible, it is uncomfortable and difficult. This is because the arm motions complement the movement of the legs and provide a little extra thrust.

As you already know, the arm movement involves coordinating the motion of both arms forward and rearward in a sequence opposite the movement of the legs. In other words, when the right leg moves forward, the right arm moves toward the rear, in the opposite direction. Conversely, because each arm moves opposite to the other, your left arm moves in the same direction as your right leg, and vice-versa. This alternating, crossing action of the upper and lower extremities constitutes the main balancing element in running. If you try to run by advancing the leg and the arm on the same side, thus reversing the cycle, you see that the action becomes complicated, clumsy, unnatural, and very inefficient. On the other hand, when the arm motions are performed correctly, they contribute slightly to the thrust, which surely leads to greater efficiency in sprinting events and delays the onset of fatigue in long-distance runs.

As we have already indicated, the arm motions are used continually and in alternation, but these are not the only elements you need to consider to develop good arm technique. The elbows need to remain bent around 90° during the entire cycle, and their movement should be limited. That way, in the retropulsion of the shoulder, when the arm moves rearward with respect to the body, the hand will reach the level of the hip and no farther. In the opposite phase, when the arm moves forward, the movement is even shorter, such that the arm scarcely goes past the cephalocaudal, or vertical, axis of the body.

We have already said that there are no two runners who use precisely the same form, and certainly there are no two running events in which the right technique is exactly the same. For example, let's consider a 100-meter sprinter and a marathoner. In observing the running form of each one, we see instantly that even though both are running athletes, their styles are very different. The length of stride and the cadence are different, as is the twisting of the upper body. This does not mean that one of them is not using the right technique but, rather, that each discipline requires adaptations of the basic form. This criterion is valid for

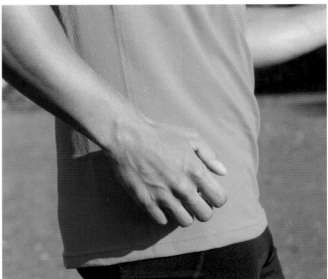

The elbows need to be bent at 90° and the hands must be relaxed.

arm movement in running. The length of stride is greater among sprinters, as is the movement of the arms, which reach farther back in the shoulder retropulsion and farther forward in the antepulsion.

To continue the analysis of arm movement, the hands must be in a relaxed position, with the fingers neither totally straight nor totally clenched in a fist. A relaxed position requires eliminating excess tension, but this certainly does not mean that the wrists and fingers are inert, because there is nothing more inefficient than moving in a "flabby" manner. We also need to consider that the hands must never cross the front of the body while moving the arms. We must move forward, so the thrust that the arms give us must be

The arms must not cross in front of the body.

Running should be as linear as possible, without raising or lowering the center of gravity.

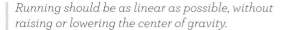

in that direction. If we cross our arms in front of our bodies in such a way that our hands are moving diagonally, we will neither produce forward thrust nor compensate effectively for the movement of our legs to produce a good dynamic equilibrium.

The work done by the abdominal and lumbar muscles is also important in stabilizing the body. This transition area

between the movements of the lower body and the arms acts like a hinge, making it possible to synchronize the dynamics of both areas, avoiding excessive impact in the process, and maintaining the upper body perpendicular to the ground or inclined slightly forward, which is one distinctive element in proper running technique.

There are some other elements that you need to keep in mind to create a technique and a style that contribute to optimum performance. First of all, the running motion should be as linear as possible. You do not run by bobbing up and down, but rather by moving forward, so most of the energy you use in vertical displacement will be wasted. All runners move their centers of gravity up and down to a certain extent, because this is a natural consequence of alternating between the aerial and the ground phases. However, all vertical movement amounts to a waste of energy that robs resources from horizontal movement, which allows us to move forward, so it must be minimized. It's a different situation if we run to burn calories; in that case, the more unnecessary movements you do, the closer you come to your goal, as in the case of casual runners who look more like they are bounding than running. In any case, there are ways to burn calories that are less harmful to the knees than bounding as you run.

Finally, if you take running to be a good way to improve health, have fun, and enjoy athletics, with no other ambition than having a good time and improving yourself, you always need to be careful of new things. Running technique and equipment have been studied for a long time; and, even though some small improvements that you can incorporate come up every year, there are not many revolutionary inventions, so you are wise to adopt a certain amount of skepticism and a critical mind in analyzing the relevant advertising.

Use proven equipment and remember to replace your running shoes regularly, because the foam intended to absorb the series of shocks from running loses its properties with use. Above all, use common sense, make the most of what people have learned in this pursuit, and interpret your sensations correctly: your body will not deceive you.

Moving vertically instead of straight ahead significantly detracts from the economy and efficiency of running.

The Benefits of Stretching for Running

Stretching exercises can bring benefits to athletes in any sport, even if they are done independently of any other physical or athletic activity. But make no mistake: stretching exercises done improperly or at the wrong time can be as harmful as any other improper activity, despite the benign aura that surrounds them.

Instinctively, we have a feeling that flexibility work will improve our form, performance, and health, no matter when, where, or how we do it. This is what we learned in school, from trainers, or from fitness coaches at the gym.

Well, even though flexibility work is highly beneficial, you need to know that it must always be done when the muscles are warmed up. Otherwise, the muscles are less elastic, and you may suffer an injury. If you include stretching exercises in your warm-up routine, they should not come at the beginning, but rather at the end or should be dovetailed with low- or medium-intensity muscle work. If you do isolated sessions of stretching, it is useful to move the area to be stretched for a few seconds or minutes before starting with the flexibility work. And, of course, if you don't have much time before heading off to compete or run, you must always give priority to a warm-up.

It certainly is not the intent for these statements to be taken as an indictment against stretching. They are simply to caution readers about the risks of improper behavior. Now we will see the many positive effects that proper behavior can provide.

In the first place, dynamic stretches included in a warm-up session can increase your mobility, which will help you improve your performance. Dynamic stretches can also be a type of warm-up and can thus contribute to the readiness of your muscles, although they must be preceded by some other warm-up exercise, such as a very gentle run. Take, for example, a skipping exercise: it may appear to be nothing more than a warm-up exercise, but it involves a chain of successive submaximum hip bends and extensions, so it performs the double function of stretching and warm-up. This also applies to touching your heels to your gluteals with every stride as you run, by extending your hips and bending your knees. Shouldn't this stretch the quadriceps? If you analyze other techniques or warm-up exercises for running, you will surely find elements of dynamic stretching. Remember that they should never be extreme or sustained movements if you do them prior to athletic exertion. But, as in all things, there are exceptions to the norm. Some athletes in sports where flexibility and amplitude of movement are essential elements in applying technique may do maximum and static stretches before athletic exertion, as in artistic gymnastics or synchronized swimming.

During athletic practice, especially if it is very prolonged, you may detect a muscle that's subject to strain, or

Including dynamic stretches in the warm-up is helpful for all athletes.

Exercises, such as skipping in any of its variants, can be done without even having to interrupt the warm-up run.

warning signs prior to a cramp, if not a cramp itself. This is common not just in running, but also in cycling, soccer, or tennis matches that seem to go on forever, such as the match that pitted Nicolas Mahut against John Isner in the 2010 Wimbledon Championship, which went on for more than eleven hours and had to be played on three different days, because of the impossibility of finishing it in a single session.

Often, in order to keep up a very prolonged athletic effort when you feel some muscle discomfort, and leaving the extreme cases for professional care, you can take some small breaks to stretch the affected muscles and massage them. This gentle stretching, combined with a massage, helps relax the muscles, thereby facilitating a return to the prior condition, and restoring blood to the area, which, on the one hand, helps with the elimination of waste products from muscle metabolism and, on the other, encourages the arrival of fresh blood, which brings nutrients and oxygen to the tissues.

Obviously, if the discomfort is constant and it stems from considerable exertion, the best course is to stop the activity immediately; however, if you are resolved to continue, because this is a personal challenge or a competition, you can find relief and improvement with stretching and a massage. The massage is intended to relax the muscles and encourage venous return, so it should be done in a direction from distal to proximal. So, if you experience strain in your gastrocnemius, you need to apply pressure from the ankle toward the knee, never in the inverse direction.

The reasons for stretching after athletic exertion are very similar to the ones already mentioned, and so is the technique to be used. In this case, the muscle is tired, because it was subjected to exertion, and you need to apply gentle stretching exercises to contribute to its relaxation and to the renewal of the blood in the area. This will encourage prompt and complete recovery.

Finally, if you want to do a flexibility session unconnected to immediate athletic exertion, you can use static stretching

Submaximal static stretching exercises after athletic performance contribute to a runner's recovery.

exercises of various types, including ones involving propioreceptive neuromuscular facilitation.

In this case, the stretching exercises are not designed for recovery, but for maintaining an adequate or optimal range of movement. Stretching can compensate for imbalances in posture and slow the loss of mobility that appears with the passage of time.

In addition, it can reduce or eliminate pain due to muscle strain or due to holding forced postures, which you often maintain as a result of your activity at work. Take, for example, working at a computer or a desk: very likely the trapezius muscle remains for hours under light but constant tension that becomes forced, because of the need to manipulate things on the desk with your hands, whether writing correspondence or using a keyboard. If the trapezius is subjected to constant tension for hours, why not stretch it? Do you suppose that gymnasts who use their muscles for hours don't do stretching exercises?

Isolated stretching sessions contribute to improving your posture and range of movement, as well as to soothing pain and tight muscles, to reducing muscle imbalances, and what's more, to improving your athletic performance.

Repetitive movements, as in running, contribute to the strengthening of the muscles; but, if these movements do not cover the entire range of movement, as in running, you may experience a shortening of this same musculature at the same time it is strengthened, so it becomes tighter every day.

And, once this stiffness sets in, what will happen if, one day, you have to exceed your normal range of movement, because you step on a rock, encounter some obstacle, or slip? It's very likely that this movement beyond the usual, greater than the one used in an ordinary stride, will produce an injury in a muscle without any slack in its range of motion.

So, oftentimes good flexibility can prevent injuries. As a result of the same process, if you have to exert yourself a little more—to "pour oil on the fire," to jump a little higher, to take a slightly longer stride, or to run a bit faster in order to win a sprint against a close competitor or to catch a bus before it leaves the stop—you surely will have a better chance of doing it if your body has a little extra range of movement. This principle applies to other traits as well: if you have more strength or endurance than you commonly use, you can deal with unforeseen events with a greater chance of success.

Extreme events or movements require a reserve capacity, including flexibility, which we don't use every day.

DYNAMIC WARM-UP STRETCHES FOR RUNNING

THE BASICS OF DYNAMIC STRETCHING

Dynamic stretches are still a major mystery for many coaches, physical trainers, and professional athletes. For many years, they have been lumped together with ballistic stretching and the risk of injury that this involves. As a result, they have been shunted aside, remaining part of only a few professional sports, such as sprinting, throwing events, and martial arts.

Nevertheless, the most recent studies indicate that, at certain times, or if they are done with specific goals in mind, dynamic stretches have many advantages over static ones. Dynamic stretching exercises help achieve a good range of movement in the joints involved, without detriment to strength or explosiveness. So, they help achieve a greater level of performance at the same time that the muscle activity contributes to the warm-up.

Until a short while ago, the general belief was that static exercises included in a warm-up routine offered many beneficial effects, such as improved performance, reduced probability of injury, and even shorter recovery time from muscle pain after exercising. Trainers, coaches, and even physical education teachers and professors took these benefits for granted, as did most athletes. The ones who are involved in any degree with running still use static stretches during their warm-ups prior to athletic activity or competition.

Unfortunately, based on the most recent scientific data available, everything seems to indicate that static stretching prior to athletic exertion does not reduce the risk of injury, and, what's worse, it in fact significantly detracts from the economy of running, especially among runners who need lots of explosiveness, such as sprinters, relay runners, hurdlers, and the like.

Various high jumping experiments have shown that individuals who do static stretches before jumping do not perform as well as those who do not stretch, and they perform far below those who do a dynamic warm-up routine.

On the other hand, a warm-up is an essential activity for any athlete, especially for those whose sports require speed, power, or great strength. A warm-up dramatically improves subsequent performance, and it has been clearly demonstrated that it reduces the risk of injury during athletic participation.

At this point, you may conclude that static stretching before exercise does not appear to be the most appropriate course for runners, and that a warm-up that involves dynamism and movement helps raise the muscle temperature and improve the irrigation and oxygenation of the muscles.

Many of the exercises that we associate with warm-ups are in fact dynamic stretches.

*Static stretches
prior to athletic
performance can be
counterproductive.*

If you closely analyze dynamic stretches, you see that they include elements useful to runners. The most important factor is the dynamic one, which contributes to the warm-up. Most of us have probably seen a runner doing high skipping before competing. This exercise is nothing more than a dynamic way of performing a broad movement of the hip joint, and thus stretching the extensor muscles of that joint, as well as some other muscles.

Thus, the benefits of dynamic stretches before athletic participation are clear. But you need to observe certain guidelines on how to do them to produce the maximum benefit and avoid possible injuries due to incorrect performance. Follow these guidelines:

■ Dynamic stretching is a series of interconnected stretching exercises that are repeated in succession and without stop until reaching the desired duration.

■ The movement used to stretch a muscle group needs to have a certain level of momentum or velocity, but without suddenness or lack of control at any time, because that would involve a risk to the joints, as well as the tendons, muscles, and ligaments.

■ At the end of the stretching movement, there should be a little bit of rebound that allows for the breadth of movement necessary to improve flexibility.

■ The duration of a set is applied to all cycles of movement, not to a static posture.

Finally, it is appropriate to recall that dynamic stretches should be part of a complete warm-up, which should include, among other things, a gentle run that can incorporate some of the chosen stretching exercises.

The Windmill

START
Stand up and place one hand over the opposite pectoral. The free arm is straight and pointing forward and down. Keep your upper body perpendicular to the floor, your back straight, and your feet about shoulder-width apart.

TECHNIQUE
Move your straight arm in a circle from front to rear, or in the opposite direction, reaching the maximum range of movement in the shoulder while still keeping your elbow straight.

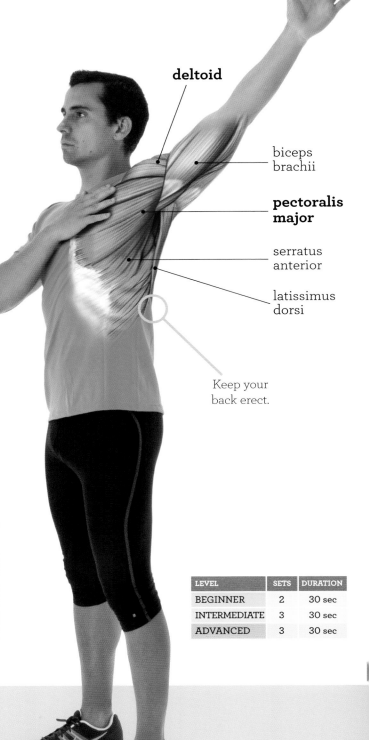

deltoid

biceps brachii

pectoralis major

serratus anterior

latissimus dorsi

Keep your back erect.

Movement Sequence

LEVEL	SETS	DURATION
BEGINNER	2	30 sec
INTERMEDIATE	3	30 sec
ADVANCED	3	30 sec

CAUTION
Perform the movement in a controlled manner to avoid momentum and sudden movements that could endanger the shoulder joint.

INDICATION
For all types of athletes, whether runners or not, and especially for those who play racquet sports, handball, and other sports that involve the upper limbs and shoulders.

Side Bends with Stride

START

Stand with one foot slightly ahead of the other. The hand on the side of the forward foot rests on your waist, and the other hand is raised above your head. Bend your upper body toward the side of the forward foot.

TECHNIQUE

Lower the raised hand and return your upper body to a perpendicular position with the floor. Take one step forward with your rear leg and raise the hand that was resting on your waist, so that both hands change positions from the preceding phase. Repeat the cycle on alternating sides with every step you take.

latissimus dorsi

tensor fasciae latae

gluteus medius

gluteus minimus

Place one foot forward in alignment with the other.

Movement Sequence

LEVEL	SETS	DURATION
BEGINNER	3	30 sec
INTERMEDIATE	4	30 sec
ADVANCED	5	30 sec

CAUTION

Do the stretches in succession, as if this were a cycle of steps, continually bending toward alternating sides and keeping up a steady rhythm so that the stretch remains dynamic.

INDICATION

For all types of athletes, including runners and walkers.

Crossed Arc

START
Place one foot slightly ahead of the other and relax your arms at your sides. Your upper body should remain erect and perpendicular to the floor.

TECHNIQUE
Extend the hip of the trailing leg as much as possible while performing maximum antepulsion with the opposite shoulder, and extend your spine. Return to the starting position and go beyond it by taking a step forward, thereby advancing the rear foot in the previous cycle. Repeat the movement, but this time extend the opposite hip and shoulder.

LEVEL	SETS	DURATION
BEGINNER	3	30 sec
INTERMEDIATE	4	30 sec
ADVANCED	4	30 sec

Movement Sequence

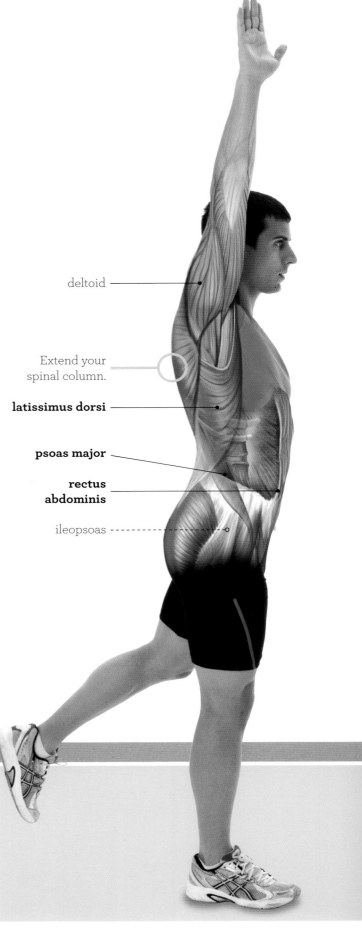

deltoid

Extend your spinal column.

latissimus dorsi

psoas major

rectus abdominis

ileopsoas

CAUTION
Do the cycle repeatedly on alternating sides, keeping up a constant speed that allows you to control the movement, but not so slowly that the dynamic feature of the stretch is lost.

INDICATION
For all types of athletes, not just runners, especially swimmers, cyclists, and triathletes.

Scissors

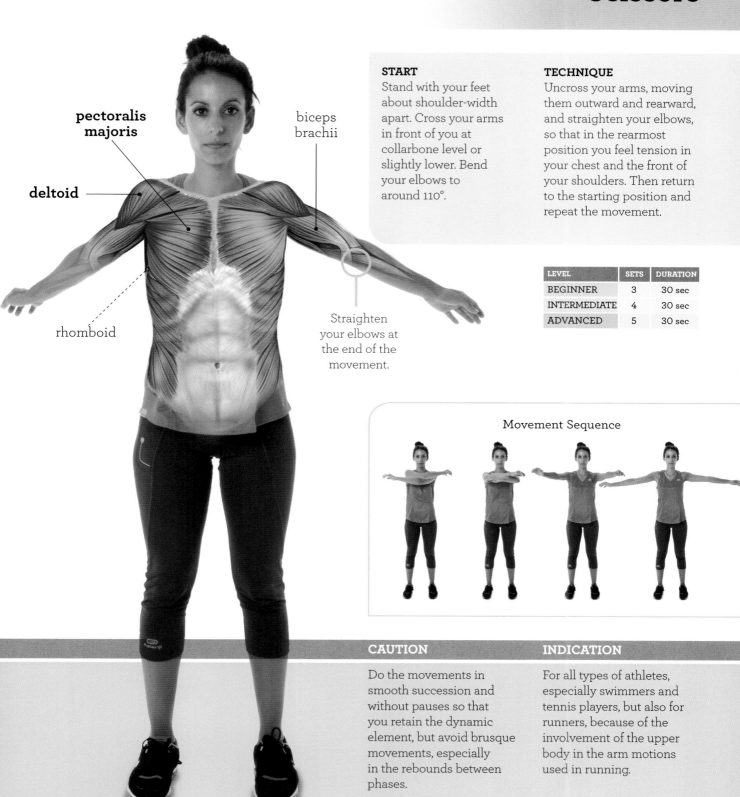

pectoralis majoris

deltoid

biceps brachii

rhomboid

Straighten your elbows at the end of the movement.

START
Stand with your feet about shoulder-width apart. Cross your arms in front of you at collarbone level or slightly lower. Bend your elbows to around 110°.

TECHNIQUE
Uncross your arms, moving them outward and rearward, and straighten your elbows, so that in the rearmost position you feel tension in your chest and the front of your shoulders. Then return to the starting position and repeat the movement.

LEVEL	SETS	DURATION
BEGINNER	3	30 sec
INTERMEDIATE	4	30 sec
ADVANCED	5	30 sec

Movement Sequence

CAUTION
Do the movements in smooth succession and without pauses so that you retain the dynamic element, but avoid brusque movements, especially in the rebounds between phases.

INDICATION
For all types of athletes, especially swimmers and tennis players, but also for runners, because of the involvement of the upper body in the arm motions used in running.

Alternating Arm Movement

START
Stand with your feet about shoulder-width apart, your back erect, and your knees straight. Hold both arms in front of you with your hands open and your elbows straight.

TECHNIQUE
Simultaneously perform antepulsion with one shoulder and retropulsion with the other, so that they move in opposite directions. When the shoulders reach their maximum range of movement, move them in the other direction, so that the raised hand lowers and the lower one rises.

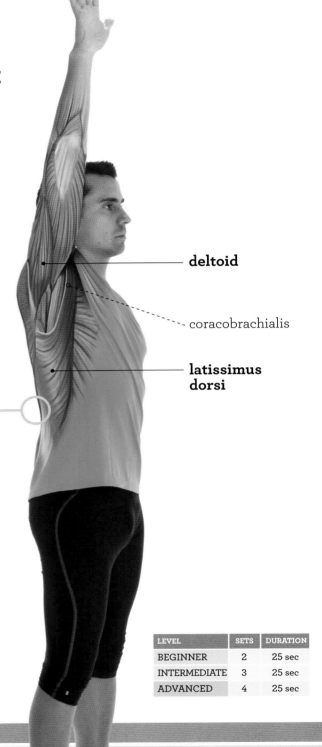

deltoid

coracobrachialis

latissimus dorsi

Keep your body erect.

Movement Sequence

LEVEL	SETS	DURATION
BEGINNER	2	25 sec
INTERMEDIATE	3	25 sec
ADVANCED	4	25 sec

CAUTION
Repeat the cycle without stopping, so that the movement is nearly continuous and the exercise is dynamic. Avoid sudden or rapid movements that could endanger your shoulders.

INDICATION
For all types of athletes, whether runners, swimmers, or others. Especially for those whose sports require the particular involvement of the shoulder and upper extremities.

Side Bend

LEVEL	SETS	DURATION
BEGINNER	3	25 sec
INTERMEDIATE	3	30 sec
ADVANCED	3	35 sec

Reach as far as possible with your hand.

latissimus dorsi

external obliques

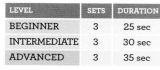

gluteus minimus

gluteus medius

tensor fasciae latae

START
Place your feet slightly farther apart than shoulder-width and keep your knees straight. Raise one hand using shoulder antepulsion, so that your fingers point to the sky and your elbow is nearly straight.

TECHNIQUE
Bend your upper body to the side as you shift your weight toward the opposite leg, in the direction of the bend. Now, the forearm of the raised hand is nearly perpendicular to the floor, and you will feel tension along the uppermost side. When you reach the maximum bend, you can bounce two or three times.

Movement Sequence

CAUTION
Lean toward alternating sides in continuous succession to preserve the dynamic element; and, if you choose to bounce at the end, always remember to avoid brusque or extreme movements.

INDICATION
For all types of athletes, including runners and players of team sports. Especially for swimmers.

Hip Rotations

START

Stand with your feet slightly apart, your upper body erect, and your hands on your hips. At this point, both your hips and your knees should be straight.

TECHNIQUE

Bend one hip and knee on the same side, so that you have just one support point. Perform abduction with this knee and lower it without touching the floor. Repeat the circular movement several times before doing it with the hip and knee of the other side.

LEVEL	SETS	DURATION
BEGINNER	2	25 sec
INTERMEDIATE	3	25 sec
ADVANCED	3	30 sec

Movement Sequence

adductor magnus

Bend your knee at the top of the movement.

gracilis

adductor minimus

adductor medius

CAUTION

Avoid brusque movements and jerking, and keep your balance throughout.

INDICATION

Especially for hurdlers, sprinters, and obstacle course runners. Also for people who play individual sports that involve running, especially ones that entail stops, fast starts, and sudden changes of direction.

Butterfly

adductor
medius

adductor
magnus

gracilis

adductor
minimus

Bring the soles of
your feet together.

START
Sit on the floor or on a
mat. Bend your hips and
knees, and bring the soles
of your feet into contact
with one another. Hold
the toes of both feet to
keep them together.

TECHNIQUE
Lower your knees and try
to move them toward the
floor simultaneously, and
then return to the starting
position. Repeat the cycle
several times, similar to
the way a butterfly flutters
its wings.

Movement Sequence

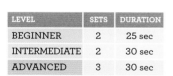

LEVEL	SETS	DURATION
BEGINNER	2	25 sec
INTERMEDIATE	2	30 sec
ADVANCED	3	30 sec

CAUTION
Do the cycles continuously,
without stopping, with
gentle bounces at the lowest
part of the movement.

INDICATION
Especially for hurdlers,
runners in obstacle events,
and sprinters. Also for soccer
players, tennis players, and
people who play sports
involving quick changes of
direction while running.

Stride and Twist

START

Stand with both arms to one side to create momentum once the exercise begins. Slightly advance the opposite foot, which you will use to take the first stride.

TECHNIQUE

Take a long stride with the leading foot, so that your center of gravity moves significantly forward and downward. At the same time, move both arms toward the side opposite the one for the starting position, accompanying the movement with a rotation of your upper body. Repeat the stride with the trailing foot, and move your arms and upper body in the opposite direction. Keep up the series of strides until you reach the specified duration.

Rotate your upper body toward the side of the forward leg.

external obliques

internal obliques

psoas major

ileopsoas

LEVEL	SETS	DURATION
BEGINNER	2	30 sec
INTERMEDIATE	3	30 sec
ADVANCED	3	35 sec

Movement Sequence

CAUTION

Create some momentum in the upper body movements, but without jerking, because you could hurt yourself or destroy the position.

INDICATION

For runners, because of the role of the abdominal muscles in stabilizing the upper body, and to a lesser degree, in breathing.

Upper Body Rotation

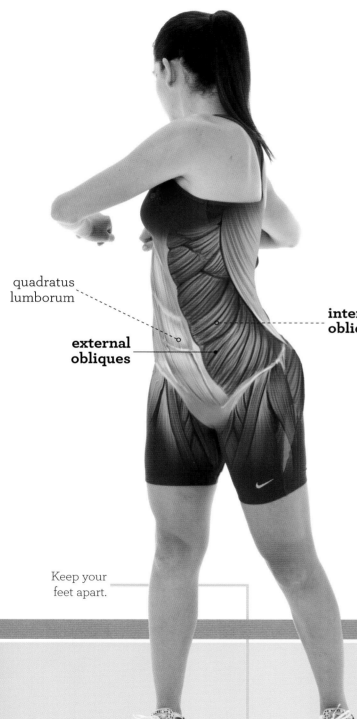

quadratus
lumborum

**internal
obliques**

**external
obliques**

Keep your
feet apart.

START

Stand with your feet no closer together than shoulder-width. Hold your hands in front of you, with your fists clenched and held together. Your elbows will be bent to about 100°. Rotate your upper body as if you wanted to see something behind you, without moving your feet.

TECHNIQUE

Rotate your upper body in the opposite direction, moving your fists rearward on the side opposite the starting one. Then return to the starting position and repeat the cycle without stops or interruptions, linking the movements together with gentle bounces between one movement and its opposite.

LEVEL	SETS	DURATION
BEGINNER	2	20 sec
INTERMEDIATE	2	25 sec
ADVANCED	2	30 sec

Movement Sequence

CAUTION

Avoid jerky bounces and keep your movements well under control.

INDICATION

For runners, swimmers, tennis players, handball players, and martial arts practitioners, and, to a lesser extent, for athletes in general, because of the importance of the abdominal muscles in stabilizing the upper body.

Leg Swings

START
Lift one foot from the floor, move it forward, and, by means of hip adduction, cross it in front of the support leg. Keep your support leg almost completely straight and rest your hands on your hips to facilitate the exercise.

TECHNIQUE
Using abduction of the forward hip, move your foot as far as possible from the medial line of your body. Then do the opposite action, so that your legs cross again. Repeat the swing, linking several movement cycles together.

Keep your upper body as perpendicular to the floor as possible.

adductor minimus

adductor medius

adductor magnus

gluteus minimus

gluteus medius

LEVEL	SETS	DURATION
BEGINNER	2	20 sec
INTERMEDIATE	2	25 sec
ADVANCED	2	30 sec

CAUTION
Avoid stopping during the exercise, and move your hip as far as possible without resorting to jerky movements or sudden stops.

INDICATION
For hurdlers, for runners on obstacle courses, and for athletes in sports involving direction changes, quick starts, and sudden stops—that is, soccer, indoor soccer, wrestling, tennis, basketball, and so on.

Movement Sequence

Slalom

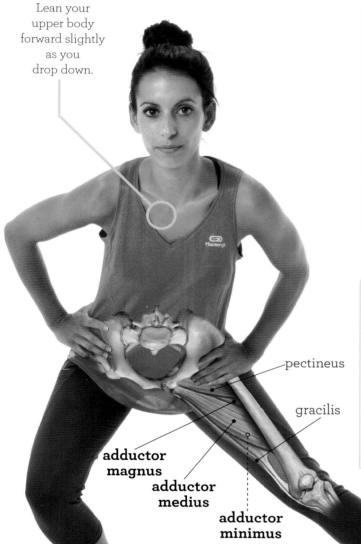

Lean your upper body forward slightly as you drop down.

pectineus

gracilis

adductor magnus

adductor medius

adductor minimus

START
Place your feet far apart, about twice your shoulder-width, with your toes pointing straight ahead. Keep your knees straight and your hands on your hips. Your upper body will be perpendicular to the floor and in line with the midpoint between your feet.

TECHNIQUE
Move your upper body to one side while bending the knee on the same side and keeping the knee of the opposite leg straight, until you reach the stretching position. Reverse the movement, passing through the starting position and arriving at the end position on the other side. Repeat the cycle several times.

Movement Sequence

LEVEL	SETS	DURATION
BEGINNER	2	20 sec
INTERMEDIATE	2	25 sec
ADVANCED	2	30 sec

CAUTION
Avoid very quick movements and abrupt stops or bounces.

INDICATION
For runners, especially hurdlers, and for athletes who take part in obstacle events or sports involving running, direction changes, and sudden stops and starts.

Ankle Rolls

START

Stand with one foot flat on the floor and the other one on tiptoes. The foot flat on the floor will support most of the body weight, and the other one will support very little weight. You can put your hands on your hips or let them hang relaxed on both sides of your body.

TECHNIQUE

Rotate the ankle that bears little weight, so that your toes remain in contact with the floor. Do several complete rotations with each leg to stretch the muscles in the forward, middle, and outside areas of your leg.

LEVEL	SETS	DURATION
BEGINNER	2	20 sec
INTERMEDIATE	3	20 sec
ADVANCED	3	25 sec

Movement Sequence

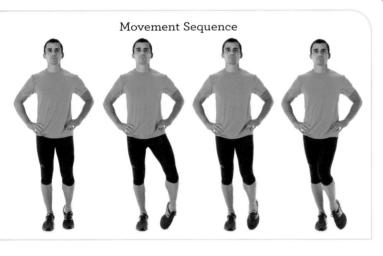

CAUTION

Don't begin the exercise until you are supported securely.

INDICATION

For all types of runners, especially those who run on uneven surfaces.

tibialis posterior

tibialis anterior

peroneus longus

peroneus brevis

Keep your toes in contact with the floor.

Assisted Hip Flex

Pull on your knee with both hands.

gluteus medius

gluteus maximus

START
Place one foot ahead of the other, at a distance of about one short step. Keep your back erect and your hands slightly in front of the axis of your body, ready to perform the stretch.

TECHNIQUE
Raise the trailing leg by bending at the hip. Hold the raised knee with both hands and pull it toward your upper body to bend the hip farther. Let go of the knee and move the corresponding foot ahead one step. Then do the same sequence of movements with the other leg, moving ahead one step each time.

LEVEL	SETS	DURATION
BEGINNER	3	20 sec
INTERMEDIATE	3	25 sec
ADVANCED	3	25 sec

Movement Sequence

CAUTION
This stretch involves no particular risk, even though you have to keep your balance in the various phases of the exercise, especially when there is only one support point.

INDICATION
For all types of runners, especially sprinters and hurdlers, and for people who take part in obstacle events, trail running, and skyrunning.

Stair Steps

START

Take a position on a step, a curb, or some other object slightly above ground level. Stand on it with the front of your feet and lean your upper body slightly forward to keep your balance.

TECHNIQUE

Lower one heel through dorsal ankle flexion at the same time as you raise the other one, and then repeat this action by reversing the role of each foot, as if you were going up a staircase. Keep up this cycle of movements without stopping.

LEVEL	SETS	DURATION
BEGINNER	3	25 sec
INTERMEDIATE	3	30 sec
ADVANCED	3	35 sec

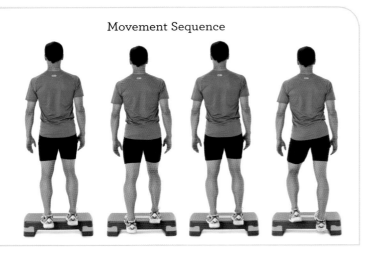

Movement Sequence

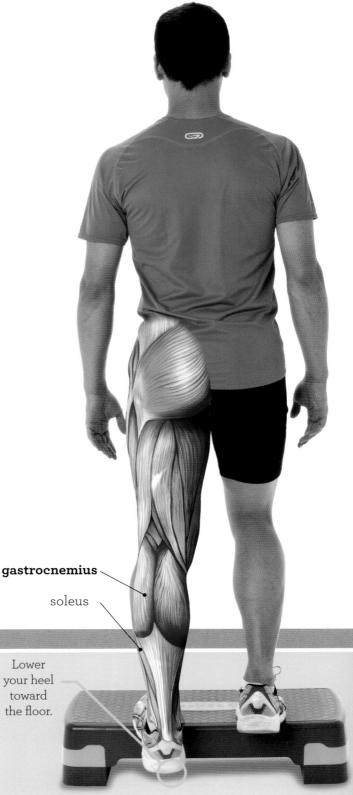

gastrocnemius

soleus

Lower your heel toward the floor.

CAUTION

Hold onto a solid or heavy object that cannot tip over during the exercise, and lean your upper body forward slightly.

INDICATION

For all types of runners, especially those who participate in long-distance events or lengthy workouts, such as marathoners, triathletes, and the like.

Goose Step

START
Place one foot ahead of the other as if you were stopped in mid-stride. Place your hands similarly, advancing the hand on the side of the trailing leg.

TECHNIQUE
Move the trailing leg forward by bending at the hip, so that the leg travels upward as far as it will go. Then lower your leg one step ahead of your starting position and do the same motion with the opposite leg. Repeat the cycle several times.

Keep your knee straight at the top of its movement.

popliteus

ischiotibials

gluteus maximus

Movement Sequence

LEVEL	SETS	DURATION
BEGINNER	2	15 sec
INTERMEDIATE	2	20 sec
ADVANCED	2	25 sec

CAUTION
Avoid excessive speed and momentum in the leg being stretched, as well as rough bounces, because these could hurt your ischiotibial muscles.

INDICATION
For all types of runners, especially sprinters and hurdlers, and for athletes who have to deal with very uneven surfaces in their events, such as skyrunners and trail runners.

STATIC STRETCHES AFTER RUNNING

THE BASICS OF STATIC STRETCHES

Static stretches are surely the most popular ones, and they have traditionally been associated with athletics. Even though it is certain, as we saw in the previous section, that they are not the best choice for warm-ups prior to a workout or a competition, except for a few sports, such as rhythmic gymnastics, they can be hugely beneficial if done at other times.

Static stretches have been shown to be more effective when the goal is to reach the greatest possible range of movement, especially with passive stretches and ones done through proprioceptive neuromuscular facilitation. They may also offer advantages at certain times related to competition. For example, if athletes feel strain or extreme fatigue in muscles during competitions, they can apply gentle stretching combined with massage on the areas, and approach the subsequent trials or games with greater certainty of getting through them, as you see on many occasions during breaks in tennis matches.

If done after the athletic exertion, especially in sports where muscle congestion during training or competition is very common—such as 4 × 400 relay races and strength and hypertrophy sports—static stretching can encourage renewal of the blood in the muscles, and thus bring a greater quantity of oxygen and nutrients to the muscles. This will help with at least partial recovery of the muscles involved.

Also, stretching sessions not connected with athletic exertion—in other words, those done in isolation—are a good way to increase flexibility and encourage relaxation. For these types of sessions, you need to remember that every stretching session must be associated with a warm-up, however brief, of the area involved, because the muscles need to be prepared before subjecting them to tensions like the ones involved in static stretching. In any case, if your sport is running, extreme flexibility could turn out to be counterproductive, because it might cause joint instability, which would not

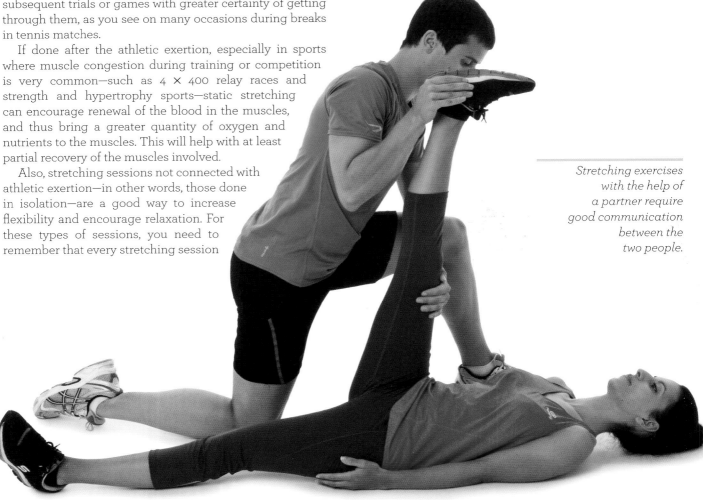

Stretching exercises with the help of a partner require good communication between the two people.

help in a sport where joint stability is essential for dealing with the repetitive impact of running, especially if the running is done on uneven surfaces.

Sometimes athletic practice is done in groups or in the company of a friend, spouse, or even a coach. This may make it possible to stretch with assistance. In other words, the companion can help during the stretch to go a little farther than if the stretch was done solo.

In theory, this is a plus. But, you have to make sure that the colleague knows how to assist and that there is constant communication. Good communication and the right speed of execution will greatly reduce the risk of injury. It goes without saying that sudden jerks and fooling around must never be done while assisting a colleague with a stretching exercise, because the results could be catastrophic.

A few things must be kept in mind while doing static stretches:

■ Static stretches with a duration between 15 and 30 seconds have been shown to be most effective in increasing flexibility, although some authors recommend durations of up to 60 seconds, especially in the case of very powerful muscle groups.

■ A stretching exercise can be done several times in a single workout, as long as you build in some good rest time between the sets.

■ Improved flexibility, like every improvement in physical activity, comes gradually, and overdoing the intensity of the stretches will not produce faster progress, but almost certainly an injury.

■ It is crucial to set priorities, because not all muscles in the body have the same degree of tightness or flexibility, and focusing on the ones that are already more flexible or neglecting the tightest ones will lead to less economy in running and a decline in athletic performance.

Finally, you must remember that stretching must be done with a warmed-up muscle group, regardless of the time chosen for the stretching session, and that the combination of warm-up and stretching exercises will always offer better results than stretches done in isolation, whether they are dynamic or static.

If we work out alone, active stretching can make it easier to stretch many muscle groups.

UPPER BODY STRETCHES

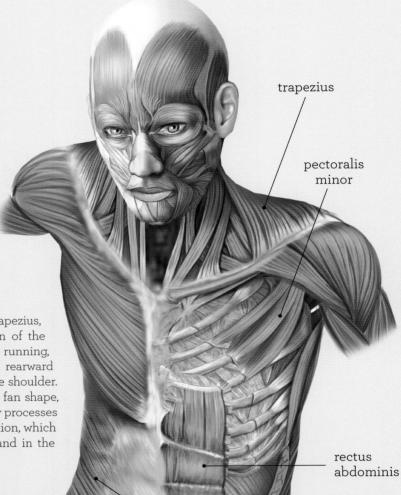

trapezius

pectoralis minor

rectus abdominis

transversus abdominis

internal oblique

external oblique

TRAPEZIUS

The upper and middle parts of the trapezius, respectively, produce elevation and adduction of the scapula, so they are used in arm movement in running, especially in the phase when the arm moves rearward relative to the body through retropulsion of the shoulder. This muscle has a broad origin because of its fan shape, which includes the occipital bone and the spiny processes of the cervical and thoracic vertebrae. Its insertion, which is much smaller, is located at the acromion and in the spine of the scapula.

LATISSIMUS DORSI

This powerful muscle performs the functions of retropulsion, adduction, and medial rotation of the shoulder. The retropulsion function comes into direct play in arm movements for running, moving the arm rearward. This muscle originates at the spiny processes of the sacral vertebrae T6 through L5 and in the posterior crest of the ilion, and it inserts at the third proximal of the humerus.

RECTUS ABDOMINIS

This muscle's most visible function is flexion of the upper body, but it also contributes to the support and protection of the internal organs, to forced exhaling, and to maintaining statically and dynamically balanced posture, so it is of great importance not only in running, but also in all other physical and athletic activities. It originates in the pubis and inserts at ribs five, six, and seven and the sternum.

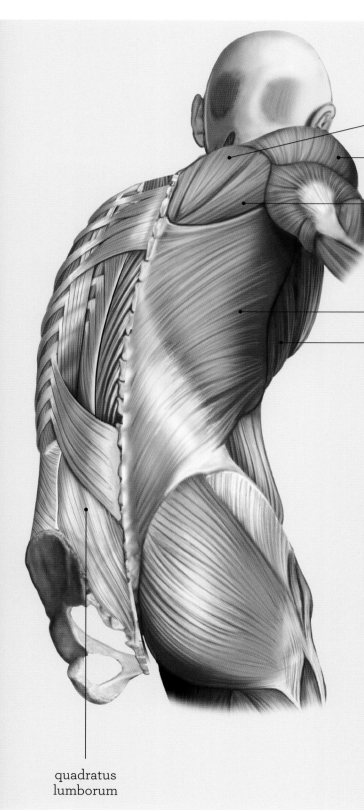

teres minor

deltoid

teres major

latissimus dorsi

serratus anterior

quadratus
lumborum

EXTERNAL AND INTERNAL OBLIQUES

Both of these muscles aid in rotating the upper body, and they work with the rectus abdominis and the transversus abdominis in internal organ support, forced exhaling, and proper static and dynamic posture, so they are extremely important in running and many other sports.

The external oblique originates at ribs five through twelve, and it inserts at the iliac crest, the thoracolumbar fascia, the linea alba abdominis, and the pubis.

The internal oblique originates at the iliac crest, the thoracolumbar fascia, and the inguinal ligament, and it inserts at ribs nine through twelve, the aponeurosis of the transversus abdominis, the inguinal ligament, the linea alba abdominis, and the cartilage of ribs seven through nine.

QUADRATUS LUMBORUM

These two muscles, one on each side of the body, produce the extension of the lumbar spinal column when they work together, and lateral flexion of the upper body if they work unilaterally. When they work at the same time, they perform the function opposite that of the rectus abdominis, so a balance between the two is largely responsible for a person's posture, whether during daily life or while running or doing other athletic activities. The quadratus lumborum originates at the crest and the internal edge of the ilion, and it inserts at the lower edge of the twelfth rib and at the transverse apophyses of vertebrae L1 through L4.

Crossed-arms Pull

START
Stand with your upper body leaning forward at less than 90° with respect to your thighs. Spread your feet shoulder-width apart and let your arms hang down.

TECHNIQUE
Cross your arms and hold the outside of your thighs just above your knees. Each hand holds the opposite leg. Then pull upward with your upper body as if you were trying to stand up, but without letting go with your hands or changing your holding points. You will feel the tension in the upper part of your back, at shoulder-blade level.

rhomboideus   **trapezius**

Keep the muscles in your upper back relaxed.

Starting Position

LEVEL	SETS	DURATION
BEGINNER	2	15 sec
INTERMEDIATE	3	25 sec
ADVANCED	4	00 sec

CAUTION
If you experience back pain this is not the best exercise for you, so you should avoid doing it, or do it very carefully, remaining very attentive to any sign of pain or discomfort.

INDICATION
For anyone, regardless of involvement in athletic activity, but especially for people who experience pain, knots, or tension in the upper part of the back.

Antepulsion with Crossed Hands

Keep your hands crossed.

serratus anterior

teres minor

latissimus dorsi

START
Keep your back totally erect, raise your arms, and cross your wrists directly above your head. When you reach this position, make sure your elbows are bent, your arms are relaxed, and your feet are far enough apart for stability while you do the exercise.

TECHNIQUE
Straighten your elbows so that your hands move upward while remaining crossed and in contact with one another. If you straighten your elbows fully while keeping your wrists crossed and perpendicular to your spine, the exercise will fulfill its purpose, but it will not be effective if even one of these three elements is lacking.

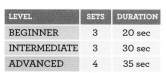

LEVEL	SETS	DURATION
BEGINNER	3	20 sec
INTERMEDIATE	3	30 sec
ADVANCED	4	35 sec

Starting Position

CAUTION
Even though there is no risk associated with this exercise, it requires maximum shoulder antepulsion, which can cause discomfort in people with prior injuries or instability in this joint.

INDICATION
For runners, because even though the latissimus dorsi rarely causes problems due to tightening or excess tension, it is used in arm movements for running. Also, especially for triathletes, because of the heavy involvement of this muscle in swimming.

Raised Arm

START
Stand erect with your feet slightly farther apart than shoulder-width. Raise one hand so that the elbow is totally straight and your fingers point skyward. You can rest the other hand on your waist for greater comfort.

TECHNIQUE
Lean the raised arm toward the opposite side as you bend the upper body slightly in the same direction. Remember that the important thing is the tilt of the arm, not how far you can move your hand. After a few seconds, repeat the stretch toward the opposite side.

LEVEL	SETS	DURATION
BEGINNER	3	20 sec
INTERMEDIATE	3	30 sec
ADVANCED	4	35 sec

Lean your arm as much as possible, involving the shoulder joint.

teres major

latissimus dorsi

external oblique

Starting Position

CAUTION
Keep your feet far enough apart to avoid instability when you lean your arm and bend your upper body to the side.

INDICATION
For all kinds of runners, especially for those who combine running with swimming, such as triathletes.

Upper Body Side Bend

teres major

latissimus dorsi

external oblique

Starting Position

Keep your feet far enough apart for stability.

START
Fold your hands so that your palms are facing outward and the backs are facing you. Raise your arms, and, with your elbows straight and your fingers interlaced, point your palms upward.

TECHNIQUE
When you bend your upper body to the side while maintaining the position of your arms and hands, you will feel tension from the stretch in the side of your back. Remember that this is a stretch on one side only, so you have to do it for each side.

LEVEL	SETS	DURATION
BEGINNER	3	20 sec
INTERMEDIATE	3	30 sec
ADVANCED	4	35 sec

CAUTION
If you feel any discomfort in your fingers while you do this exercise, relax your hand position slightly.

INDICATION
For runners in general, because of the involvement of the latissimus dorsi in arm movements for running. Especially for runners who combine running events with swimming, such as triathletes, or those who participate in other sports that involve the upper body.

Side Slip

START

Get down on all fours, holding yourself up on both hands and knees. Your knees must be perpendicular to your hips, and your hands a short distance ahead of your shoulders. Keep your elbows and your spine straight.

TECHNIQUE

Slide one hand ahead while maintaining contact with the floor. This will cause your body to tip to one side, and the side being stretched will come closer to the floor. The maximum shoulder antepulsion will create sufficient tension in the latissimus dorsi to produce the stretch.

Starting Position

LEVEL	SETS	DURATION
BEGINNER	2	20 sec
INTERMEDIATE	2	30 sec
ADVANCED	3	35 sec

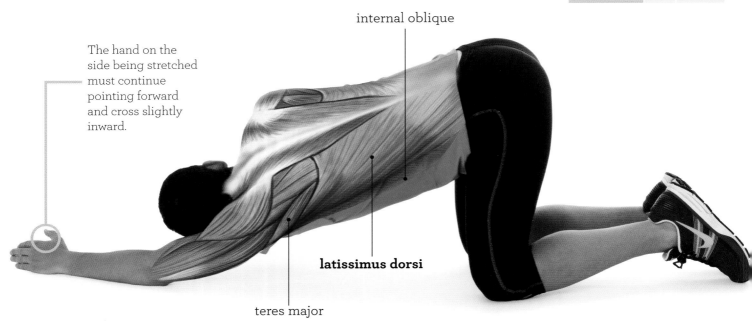

internal oblique

The hand on the side being stretched must continue pointing forward and cross slightly inward.

latissimus dorsi

teres major

CAUTION

Do the movement slowly and gradually, keeping most of the weight on the arm that remains stationary.

INDICATION

For people who experience tension in the middle and lower back and for runners, because of the involvement of the latissimus dorsi in arm movements for running, as well as for triathletes, because of this muscle's use in swimming.

Unilateral Pull with Support

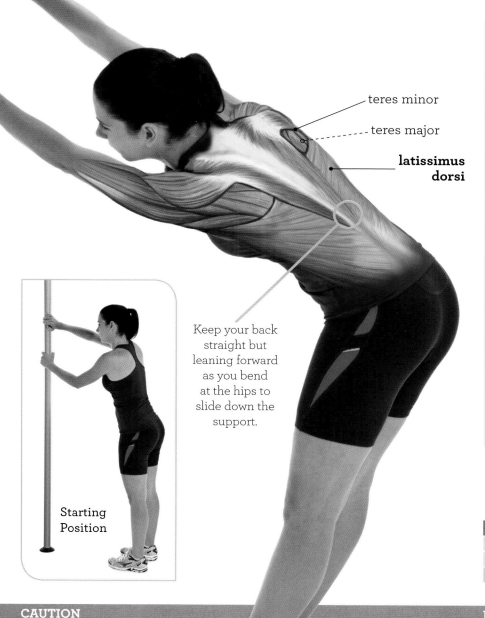

teres minor

teres major

latissimus dorsi

Keep your back straight but leaning forward as you bend at the hips to slide down the support.

Starting Position

START
Stand facing a solid vertical support that you can use for pulling. Hold onto it with both hands, one above the other, with both palms facing the same direction. Your feet should be at least a short distance apart, with your weight distributed on them evenly.

TECHNIQUE
Slide your hands downward by bending at the hips, keeping your back and your legs straight. At this point, the lower arm will push against the support and the upper one will pull on it.

LEVEL	SETS	DURATION
BEGINNER	2	20 sec
INTERMEDIATE	3	35 sec
ADVANCED	4	30 sec

CAUTION

Make sure that the support you use is sturdy enough and that both your hands and your feet are in secure contact with the support and the floor.

INDICATION

For runners, especially those who combine running with other athletic activities, such as triathletes, and for people who experience tension in the middle and lower parts of the back.

Cobra Position

START

Lie down on your stomach and place your hands near your shoulders, as if you were about to do a push-up. Your chest, abdomen, and thighs should touch the floor. Your legs and ankles should be relaxed and your elbows should be fully bent.

TECHNIQUE

Straighten your elbows, lifting your chest and abdomen, but keep your hips on the floor or very close to it, to produce an extension in your spine and a stretch in your abdominal muscles.

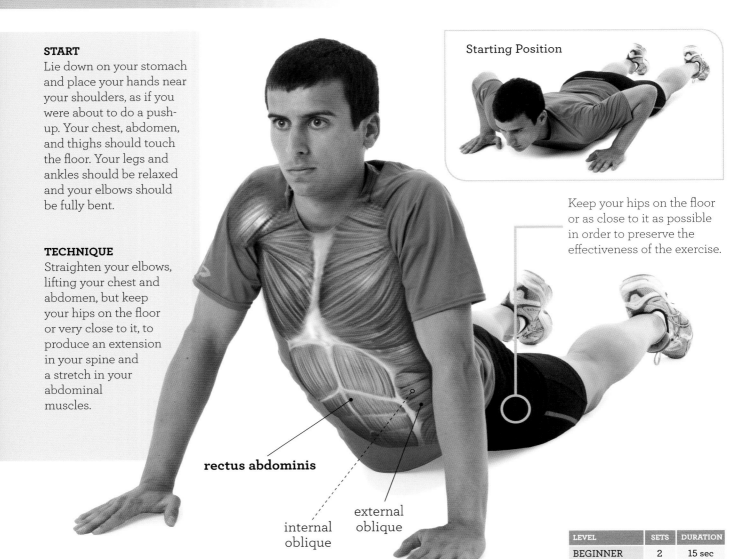

Starting Position

Keep your hips on the floor or as close to it as possible in order to preserve the effectiveness of the exercise.

rectus abdominis

internal oblique

external oblique

LEVEL	SETS	DURATION
BEGINNER	2	15 sec
INTERMEDIATE	2	20 sec
ADVANCED	3	25 sec

CAUTION

If you experience problems or pain in your back, avoid this exercise or reduce the intensity and the duration.

INDICATION

For all runners, because the abdominal muscles contribute to upper body stability. Especially for sprinters, because these muscles are used in breathing, a particularly important element in middle- and long-distance events.

Upper Body Rotation with Support

internal oblique

external oblique

quadratus lumborum

Keep your feet apart for stability.

START
Stand with your feet about twice shoulder-width apart and keep your knees straight. Bend at the hips, lean your upper body forward, and touch one of your ankles with both hands. Place your palms just above the ankle.

TECHNIQUE
Raise the hand opposite the ankle you are holding and move it upward and rearward without letting go with the other one, as if you were trying to reach something behind your back. The upper body twist that results will simultaneously stretch the external oblique muscle on one side and the internal oblique on the other.

LEVEL	SETS	DURATION
BEGINNER	2	20 sec
INTERMEDIATE	2	25 sec
ADVANCED	3	30 sec

CAUTION
Don't be too eager to stretch your abdominal muscles, because upper body balance, support for internal organs, and forced exhaling depend in large measure on their firmness.

INDICATION
For runners of all distances, because good abdominal muscle conditioning affects upper body stability and helps with proper breathing.

Starting Position

Upper Body Rotation with Pole

START

Place your feet shoulder-width or slightly farther apart. Take a pole, broom handle, dowel, or the like and hold it by both ends behind your neck, resting on your shoulders. The degree of elbow bend is not crucial, and you can even do without the pole if you don't have one available.

TECHNIQUE

Rotate your upper body in such a way that one end of the pole points to the front and the other end toward the rear. Hold the position for a few seconds and then rotate in the opposite direction. In this exercise, the external oblique on one side works simultaneously with the internal oblique on the opposite side. You can do this stretch dynamically, continually rotating in alternating directions.

Starting Position

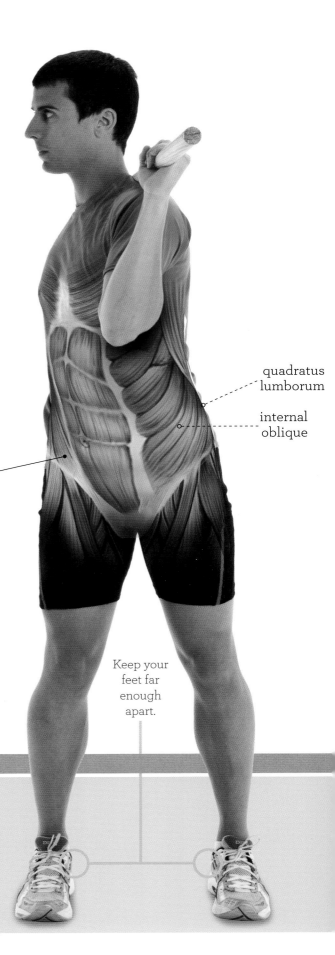

quadratus lumborum

internal oblique

external oblique

Keep your feet far enough apart.

LEVEL	SETS	DURATION
BEGINNER	2	20 sec
INTERMEDIATE	2	25 sec
ADVANCED	3	30 sec

CAUTION

If you do this exercise dynamically, avoid sudden changes of direction, but don't totally omit the little bounce involved.

INDICATION

For all types of runners, regardless of events or distances, because the abdominal muscles contribute to upper body stability and respiratory function.

Supine Upper Body Rotation

Starting Position

START

Lie down on your side, and move your upper leg forward and the lower one rearward, so that your feet remain in contact with the floor but separated from one another. The lower hand can rest on your leg or the floor near it. The uppermost arm reaches toward the front.

TECHNIQUE

Rotate your upper body so that the forward hand moves rearward over the top of you until it touches the floor behind you (if possible). Stop the movement and hold the position for a few seconds. This will stretch the external oblique on one side and the internal oblique on the other.

The forward foot must always remain in contact with the floor.

quadratus lumborum

external oblique

internal oblique

LEVEL	SETS	DURATION
BEGINNER	2	20 sec
INTERMEDIATE	2	25 sec
ADVANCED	3	30 sec

CAUTION

Do this exercise slowly, especially if you suffer from back pain or have any spinal injury, and don't lie down on very hard surfaces if you can avoid it.

INDICATION

For runners in all sports, because of the importance of the abdominal muscles in breathing and in upper body strength. Especially for triathletes.

Supine Crossover

START
Lie down on your back and hold your arms out to the side, with both palms touching the floor to act as support. Then bend your hips and knees 90°, keeping your legs together, as you would do when sitting down on a chair.

TECHNIQUE
Without changing the degree of bend in your hips and knees, rotate your legs to one side. Try to touch the floor with the outside of the lower thigh, without raising your upper back.

Starting Position

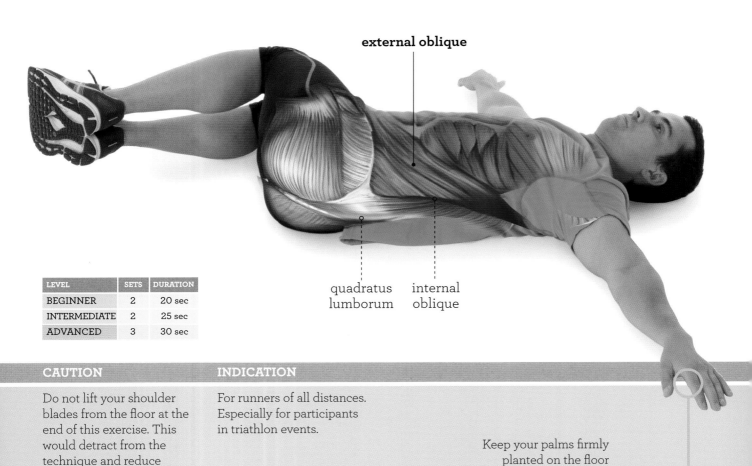

external oblique

quadratus lumborum

internal oblique

LEVEL	SETS	DURATION
BEGINNER	2	20 sec
INTERMEDIATE	2	25 sec
ADVANCED	3	30 sec

CAUTION
Do not lift your shoulder blades from the floor at the end of this exercise. This would detract from the technique and reduce the effectiveness of the exercise.

INDICATION
For runners of all distances. Especially for participants in triathlon events.

Keep your palms firmly planted on the floor for support, so the top of your back remains in contact with it.

Standing Crossover

Starting Position

internal oblique

external oblique

quadratus lumborum

Keep your arms parallel to the floor.

START
Place your feet about shoulder-width apart. Hold your arms out to the side at a 90° angle to your upper body, with your arms totally straight. Keep your back straight and perpendicular to the floor.

TECHNIQUE
Move your upper body to the side, so that your spine is no longer perpendicular to the floor and your shoulders are no longer aligned with your hips. Keep your arms out to the side and parallel to the floor. The movement is very short and, in no case, should you move your feet from their initial contact point.

LEVEL	SETS	DURATION
BEGINNING	3	20 sec
INTERMEDIATE	5	30 sec
ADVANCED	6	40 sec

CAUTION
Keep your feet far enough apart to maintain your balance in the final phase of the exercise.

INDICATION
For athletes who experience tension in the back, especially if it is due to excessive spinal curvature.

UPPER EXTREMITY, SHOULDER, AND CHEST STRETCHES

DELTOID

This muscle has three sections with different functions.

Anterior or clavicular section: Its main function is antepulstion of the shoulder, an action that results when one upper extremity is moved forward in arm movements associated with running. This improves the runner's momentum and dynamic equilibrium. This muscle originates in the third distal of the clavicle, and it inserts at the deltoid tuberosity of the humerus.

Middle or acromial section: Its main function is shoulder abduction, so it is the one least involved in arm movements for running, although it keeps the arm slightly separated from the body. Its origin is at the upper part of the acromion, and it inserts at the deltoid tuberosity of the humerus.

Posterior or spinal section: This is responsible, in part, for shoulder retropulsion, in which it is possible to move the upper extremity rearward with respect to the body through arm movements in running. This section also participates in shoulder abduction. It originates at the spine of the scapula, and it inserts at the deltoid tuberosity of the humerus.

PECTORALIS MAJOR

The functions of this muscle are antepulsion, adduction, and internal rotation of the shoulder, so it is used along with the deltoid in arm movements of running, moving the arm forward with respect to the medial line of the body and, thus, contributing to the runner's momentum and stability. It originates at the anterior surface of the clavicle, the body of the sternum, the anterior costal cartilage of ribs one thorugh six, and the aponeurosis of the oblique muscle, and it inserts at the intertubercular groove of the humerus.

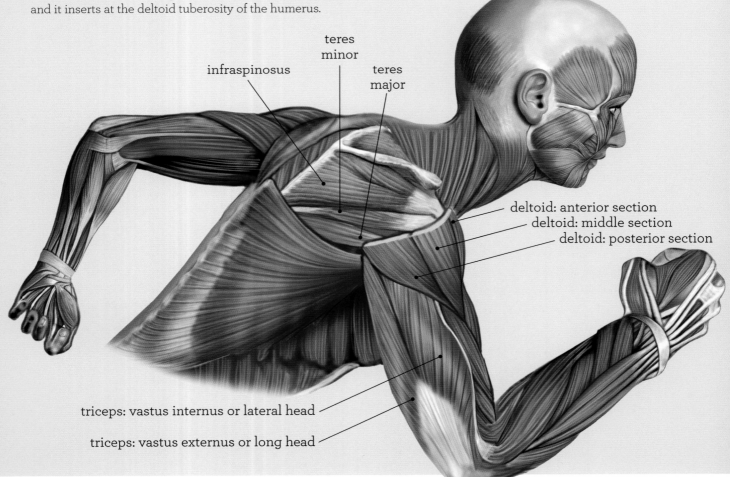

teres minor

infraspinosus

teres major

deltoid: anterior section
deltoid: middle section
deltoid: posterior section

triceps: vastus internus or lateral head

triceps: vastus externus or long head

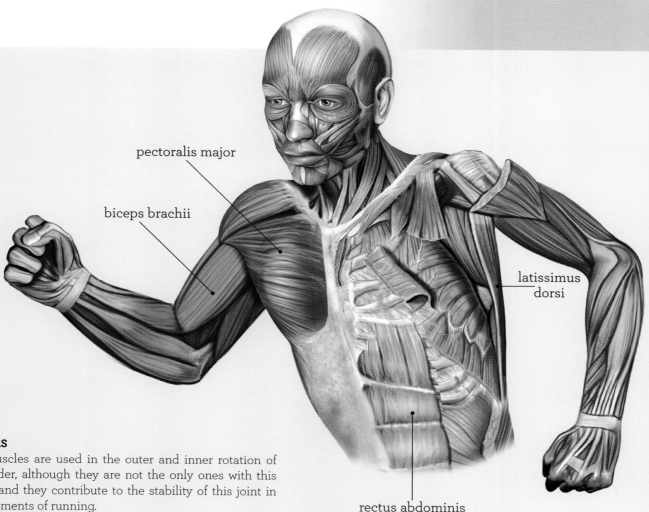

pectoralis major

biceps brachii

latissimus dorsi

rectus abdominis

ROTATORS

These muscles are used in the outer and inner rotation of the shoulder, although they are not the only ones with this function, and they contribute to the stability of this joint in arm movements of running.

Infraspinatus: This is an outer rotator of the shoulder. It originates at the infraspinous fossa of the scapula, and it inserts at the tuberculum majus of the humerus.

Teres minor: This is an outer rotator of the shoulder. It originates at the lateral edge of the scapula, and it inserts at the tuberculum majus of the humerus.

Subscapular: This is an inner rotator of the shoulder. It originates at the subscapular fossa, and it inserts at the tuberculum minus humeri.

Teres major: This is an inner rotator of the shoulder. It originates at the lower angle of the scapula, and it inserts at the intertubercular groove of the humerus.

BICEPS BRACHII

This muscle is made up of two parts, and its main function is bending the elbow and supination of the forearm, so it contributes to keeping the static bend of the elbows at around 90° in long-distance running, as well as to the continual bending and straightening of the elbows of sprinters, thereby giving runners momentum and stability. The first part of the muscle originates at the coracoid apophysis of the scapula, and it inserts at the radial tuberosity. The second part originates at the supraglenoid tuberculum of the scapula, and it inserts at the bicipital aponeurosis.

TRICEPS BRACHII

This muscle is made up of three parts, or heads, whose main functions are straightening the elbow and retropulsion of the shoulder. Therefore, they are particularly important in the rear phase of arm movements of sprinters, improving these runners' dynamic balance and momentum.

Long head: This muscle originates at the infraglenoid tuberculum of the scapula, and it shares an insertion with the other two heads at the olecranon of the ulna.

Outer head: This originates at the upper third of the humerus.

Inner head: This originates at the lower two thirds of the humerus.

Bilateral Retropulsion

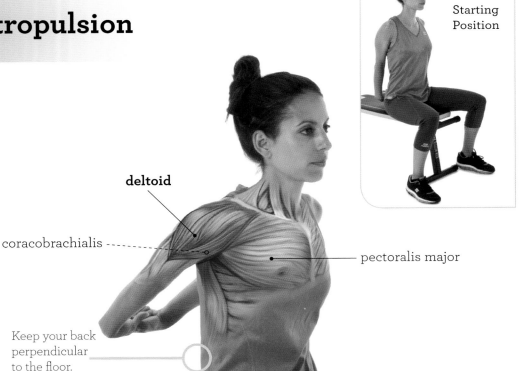

Starting Position

deltoid

coracobrachialis

pectoralis major

Keep your back perpendicular to the floor.

START

This exercise can be done either seated or standing. In either case, you will need to keep your back perpendicular to the floor, look straight ahead, and interlace your fingers behind your hips so that your palms face inward.

TECHNIQUE

Perform shoulder retropulsion as far as possible while keeping your fingers interlaced. Your back must always remain perpendicular to the floor, and you must continue to look straight ahead. This is an active stretch, so its effectiveness diminishes as the seconds go by.

LEVEL	SETS	DURATION
BEGINNER	2	10 sec
INTERMEDIATE	3	15 sec
ADVANCED	3	15 sec

CAUTION

Pay particular attention to the final position, because it is common in this type of exercise to reduce the tension and the stretch as the active muscles experience fatigue.

INDICATION

For all runners, because of the involvement of the deltoid in arm movements for running. Especially for sprinters, whose arm movements are more rapid and energetic.

Bird Position

deltoid

pectoralis major

pectoralis minor

Keep your back muscles flexed and your chest forward.

Starting Position

START
Stand with your feet in line with your shoulders and your upper body perpendicular to the floor. Keep your arms relaxed and your hands in front of you, with the palms together or very close to one another.

TECHNIQUE
Spread your arms, contract the muscles in the upper part of your back, and expand your chest, taking on a position similar to a soaring bird of prey, but staying perpendicular to the floor. Hold the maximum contraction in the upper back muscles to produce an adequate stretch in the anterior deltoid and the pectoralis major.

LEVEL	SETS	DURATION
BEGINNER	2	10 sec
INTERMEDIATE	2	15 sec
ADVANCED	3	15 sec

CAUTION
Avoid relaxing the muscles in the top of your back as the seconds go by, because the deltoid and the pectoralis major are very powerful muscles, and the effectiveness of an active stretch like this one depends on your doing so.

INDICATION
For all runners, given the involvement of the deltoid in arm movements for running. Especially for sprinters, because their movements are very forceful and require great strength and explosiveness.

Retropulsion with Support

START

Stand with your back toward a support, such as a high stool, a chair, or a bench, with one foot in front of the other. Place both hands together on top of the chosen support, with your knees slightly bent. You must keep your back perpendicular to the floor and look straight ahead.

TECHNIQUE

Lower your upper body by bending your knees, so that the shoulder retropulsion is emphasized and you feel tension from the stretch in the front part of your shoulders. Avoid bending your upper body too far forward, because this will reduce the effectiveness of the stretch.

biceps brachii

deltoid

Avoid leaning too far forward in the last part of this exercise.

pectoralis major

Starting Position

LEVEL	SETS	DURATION
BEGINNER	2	20 sec
INTERMEDIATE	3	25 sec
ADVANCED	3	30 sec

CAUTION

Stop the downward movement of your body if you feel any type of discomfort in your shoulders, but remember that a sensation of tension must be present.

INDICATION

For all runners, because of the involvement of the deltoid in arm movements for running. Especially for sprinters and triathletes, because of the special involvement of this muscle in their sports.

Assisted Rearward Pull

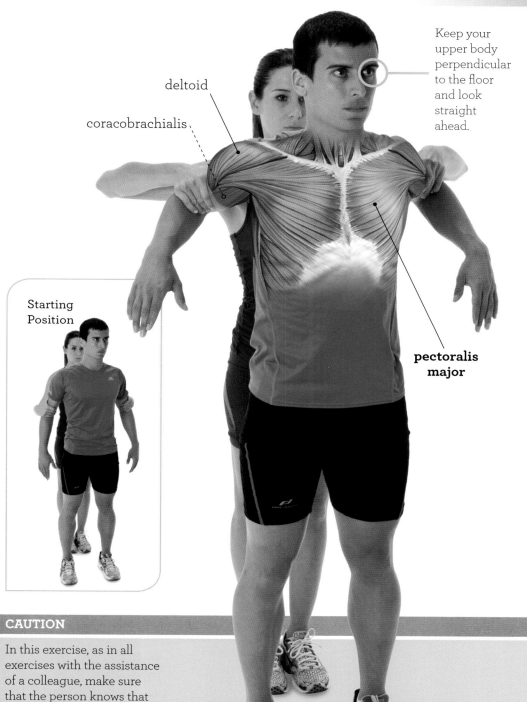

Keep your upper body perpendicular to the floor and look straight ahead.

deltoid

coracobrachialis

pectoralis major

Starting Position

START
The colleague helping you with this stretch must stand behind you a small step away and must hold your arms at the elbows while both of you remain standing and looking straight ahead. Keep your feet in line with your shoulders.

TECHNIQUE
Your colleague must pull your arms upward and rearward, thereby producing retropulsion in your shoulders and tension in your chest and the front part of your deltoid. Keep your back perpendicular to the floor throughout this exercise.

LEVEL	SETS	DURATION
BEGINNER	2	20 sec
INTERMEDIATE	3	30 sec
ADVANCED	3	40 sec

CAUTION
In this exercise, as in all exercises with the assistance of a colleague, make sure that the person knows that the pull must be applied slowly and gradually, and is alert to any signal you provide to stop once you reach the right point.

INDICATION
For all runners, because of the involvement of the pectoral muscles in arm movements for running and in breathing. Especially for people who participate in athletic events that combine several sports, such as triathlon and decathlon.

Forward Elbow Pull

START

Rest one hand on your waist, with your elbow bent around 90°, your thumb toward the rear, and your index, middle, ring, and little fingers together at the side of your abdomen. Use your free hand to hold the elbow of the other arm.

TECHNIQUE

Pull on the elbow you are holding while keeping your hand on its original support point on your waist. When you feel tension in the back part of your shoulder and shoulder blade, stop the movement and hold the position for a few seconds.

Starting Position

supraspinatus

infraspinatus

teres minor

Keep your support hand on the same spot during the entire exercise.

LEVEL	SETS	DURATION
BEGINNER	2	20 sec
INTERMEDIATE	3	30 sec
ADVANCED	3	40 sec

CAUTION

Be very careful in doing this exercise, given the instability of the shoulder and the relatively small size of the muscles being stretched.

INDICATION

For runners in general. Also for people who experience tension in the muscles of the upper back, who participate in sports using implements, who swim, or who play handball.

Assisted Retropulsion

biceps
brachii

Apply inward rotation to the arm in order to localize the stretch in the biceps brachii.

coracobrachialis

pectoralis major

START
Sit on the floor, a stool, a chair, or other similar item. Your assistant should take a position behind you and hold one of your wrists. Keep your back straight and perpendicular to the floor, with your chest and your gaze facing forward. The arm being held must be straight and relaxed.

TECHNIQUE
Your assistant pulls your wrist upward and rearward, producing shoulder retropulsion while rotating the arm inward. Your elbow must remain straight. For the exercise to be effective, you must avoid twisting your upper body.

Starting Position

LEVEL	SETS	DURATION
BEGINNER	2	20 sec
INTERMEDIATE	2	25 sec
ADVANCED	3	25 sec

CAUTION
Make sure that your assistant pulls on your wrist slowly and carefully, is alert to your instructions, and knows when to stop the movement.

INDICATION
For all runners, because this exercise relaxes the biceps brachii after the continual bending of the elbow in arm movements used in running, which necessarily involve partial, sustained contraction of this muscle.

Rearward Pull with Fingers

START

Stand or sit on a chair, a bench, or a stool. Place one arm behind your back and raise the other one, so that your forearm and hand are above your head. Keep your upper body perpendicular to the floor.

TECHNIQUE

Bend your elbows all the way and try to interlace the fingers of both hands behind your back. You may be able to interlace the fingers of both hands, or you may not even be able to touch the tips. In any case, try to go as far as possible and hold the position for a few seconds.

Raise your elbow as high as possible to maximize the stretch.

triceps brachii

teres major

latissimus dorsi

LEVEL	SETS	DURATION
BEGINNER	2	20 sec
INTERMEDIATE	2	25 sec
ADVANCED	2	30 sec

Starting Position

CAUTION

In this exercise, both shoulders are in extreme positions, so you need to be attentive to any discomfort in them and stop the movement if necessary.

INDICATION

Especially for triathletes and cyclists, because of the continuous tension to which the triceps brachii are subjected when supported on bicycle handlebars.

Rearward Elbow Pull

Bend your elbow all the way to produce a greater degree of stretching.

Starting Position

triceps brachii

teres major

latissimus dorsi

START

Sit on a bench or a stool. Raise one arm and bend your elbow to 90°, so that your forearm is parallel to the floor and behind your head. Hold the bent elbow firmly with your other hand.

TECHNIQUE

Pull rearward on the elbow you are holding, while bending it all the way. You will immediately feel the tension in the back of your arm, which will indicate that you are doing the stretch correctly. Hold the position for a few seconds and do the exercise on the other side.

LEVEL	SETS	DURATION
BEGINNER	2	20 sec
INTERMEDIATE	2	25 sec
ADVANCED	2	30 sec

CAUTION

Even though this exercise poses no significant risks, it involves forced retropulsion of the shoulder, so you need to be alert to any discomfort that may arise.

INDICATION

For all runners, because of the involvement of the triceps brachii in arm movements for running. Especially for triathletes, because of the use of this muscle in swimming and in holding onto bicycle handlebars.

HIP STRETCHES

ADDUCTORS
These are the adductor magnus, medius, and minimus. They share their main function, which is hip adduction. This makes it possible to keep the legs aligned and to prevent them from separating to an undesirable degree. The action of these muscles is particularly important not only in running sports, but also in soccer, tennis, volleyball, and judo, sports in which lateral movement is a regular element.

Adductor magnus: This muscle originates at the pubis and the ischium, and it inserts at the diaphysis and the distal epiphysis of the femur.

Adductor medius: This muscle originates at the pubis, and it inserts at the linea aspera of the femur.

Adductor minimus: This muscle also originates at the pubis, and it also inserts at the linea aspera of the femur.

PSOAS MAJOR
The major function of this muscle is to bend the hip, so it is important in the swing phase, as well as to the phase prior to the swing, in walking and running, in which the trailing leg moves ahead of the body before touching the ground again. It originates at vertebrae L1 to T12, and it inserts at the lesser trochanter of the femur.

ILEOPSOAS
The main function of this muscle, like the psoas major, is to bend the hip so that the toe muscles act in concert in the swing phase and in the phase prior to the swing in running and walking. It originates at the sacrum, the fossa, and the iliac spine, and it inserts at the lesser trochanter of the femur.

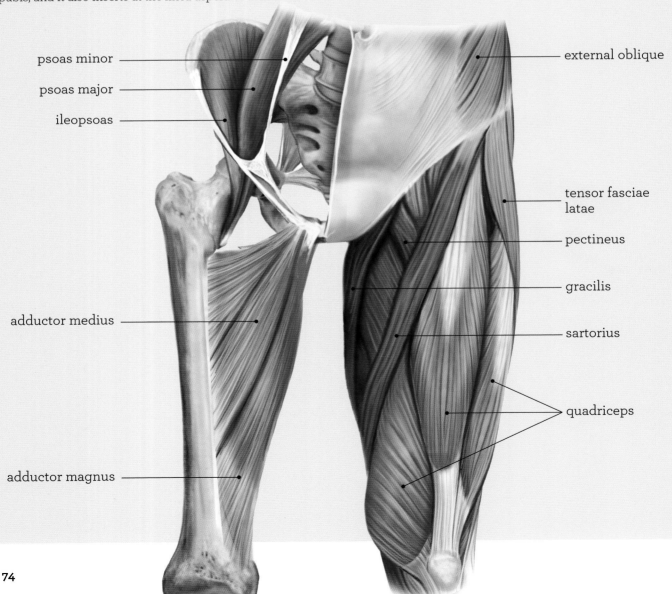

psoas minor

psoas major

ileopsoas

external oblique

tensor fasciae latae

pectineus

gracilis

adductor medius

sartorius

quadriceps

adductor magnus

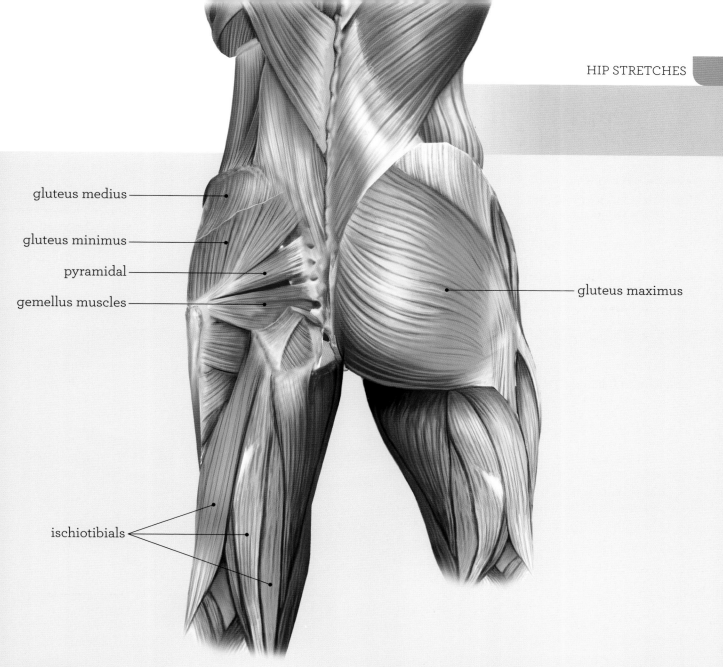

gluteus medius

gluteus minimus

pyramidal

gemellus muscles

gluteus maximus

ischiotibials

GLUTEALS

These are the gluteus maximus, medius, and minimus, and all have different functions:

Gluteus maximus: Its main function is to straighten the hip, although it also rotates outward and abducts this joint. So, it is a major player in the initial contact and support phases of running. It also provides the force necessary for the runner's forward thrust, so it becomes highly developed in sprinters. It originates at the ilium, the sacrum, and the coccyx, and it inserts at the gluteal tuberosity of the femur and in the iliotibial tract.

Gluteus medius: This muscle plays a slight role in straightening the hip. Its main function is hip abduction, so it contributes to keeping the feet aligned while running, and it plays an active part in lateral movements used in many sports, such as tennis, soccer, handball, judo, and volleyball. It originates at the posterior face of the ilium, and it inserts at the trochanter major of the femur.

Gluteus minimus: Its main function is hip abduction, although it also contributes slightly to both hip extension and flexion, so it, too, is used in lateral movements and in alignment of the lower limbs while running. It originates at the posterior face of the ilium, and it inserts at the trochanter major of the femur.

PYRAMIDAL

Its main functions are outward rotation and abduction of the hip, so its contribution to walking and running is similar to that of the gluteus medius and minimus, although of lesser importance, given its small size and modest strength. It originates at the sacrum, and it inserts at the trochanter major of the femur.

Static Butterfly

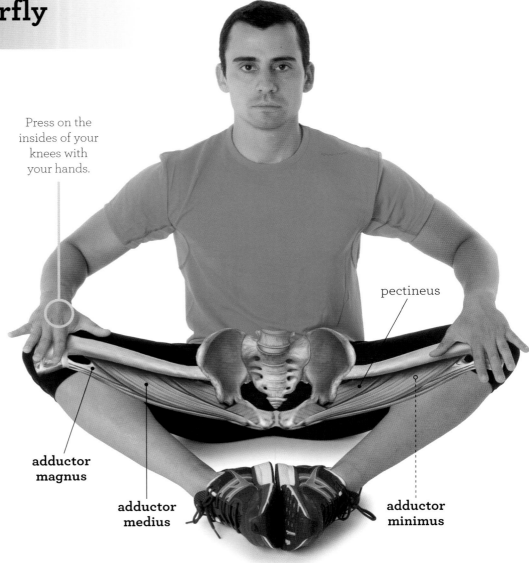

START

Sit on the floor or on a mat. Bend your knees and put the soles of your feet together, so that your feet are resting on their outer edges. Place your hands on the insides of your knees without applying pressure.

TECHNIQUE

Slowly push your knees toward the floor while keeping the soles of your feet together. You will feel a gradual increase in the tension of your adductors due to the stretch. When you reach the optimal stretching point, stop the movement and maintain the pressure on your knees for a few seconds.

Press on the insides of your knees with your hands.

pectineus

adductor magnus

adductor medius

adductor minimus

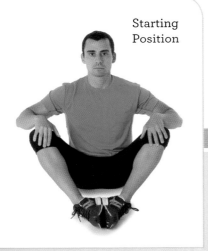

Starting Position

LEVEL	SETS	DURATION
BEGINNER	2	20 sec
INTERMEDIATE	3	25 sec
ADVANCED	3	30 sec

CAUTION

Do not shove hard or move too quickly in this stretch, because these muscles are susceptible to injury.

INDICATION

For people who do running that involves jumps or irregular terrain, such as the 3,000-meter obstacle event, hurdle events, and cross-country or trail-running events.

Hip Abduction with Support

START
Stand with your side toward a support element at least 16 inches (40 cm) high. Place the inside of the near foot onto the support and keep your knee straight. The other leg must remain firmly planted on the floor, because it will be your main support.

TECHNIQUE
Bend the knee of the support leg, so that your body gradually moves downward and the abduction of the other leg increases. You can place one hand on the leg being stretched and the other on your waist to contribute to proper execution and to your balance.

Keep the knee of the raised leg straight.

pectineus

adductor medius

adductor minimus

adductor magnus

gracilis

Starting Position

LEVEL	SETS	DURATION
BEGINNER	2	20 sec
INTERMEDIATE	3	25 sec
ADVANCED	3	30 sec

CAUTION
Be sure to keep a stable, balanced position that allows you to stop the movement at the desired point.

INDICATION
For runners in events that involve irregular terrain or involve jumping over hurdles or obstacles. Also for people who play team sports that require distance running.

Alternating Hip Abduction

START
Stand with your feet twice shoulder-width apart. Neither foot should be ahead of the other, so that you have a good support base while doing the exercise. Place your hands on your hips or waist.

TECHNIQUE
Bend one of your knees and shift your weight toward that side, producing abduction of the opposite hip. As you lower your center of gravity, lean your upper body slightly forward. This will help you keep your balance.

Lean your upper body forward.

adductor magnus

adductor medius

adductor minimus

pectinius

gracilis

LEVEL	SETS	DURATION
BEGINNER	2	20 sec
INTERMEDIATE	3	25 sec
ADVANCED	3	30 sec

Starting Position

CAUTION

Keep your feet planted firmly while you do this exercise.

INDICATION

For hurdlers, because of the need for flexibility in the trailing leg while going over the hurdle, and for cross-country and obstacle course runners, plus athletes in other sports that require running with changes of direction, stops, and quick starts, such as most team sports.

Supine Butterfly

LEVEL	SETS	DURATION
BEGINNER	2	20 sec
INTERMEDIATE	3	25 sec
ADVANCED	3	30 sec

START
Lie down on your back, bend your knees, and put the soles of your feet together. You can rest your hands on your hips or the inside of your thighs near your knees if you want to use them.

TECHNIQUE
Lower your knees so that they move farther apart and closer to the floor. If you want to increase the intensity of the stretch, you can press with your hands on the inside of your thighs, forcing them downward a little farther.

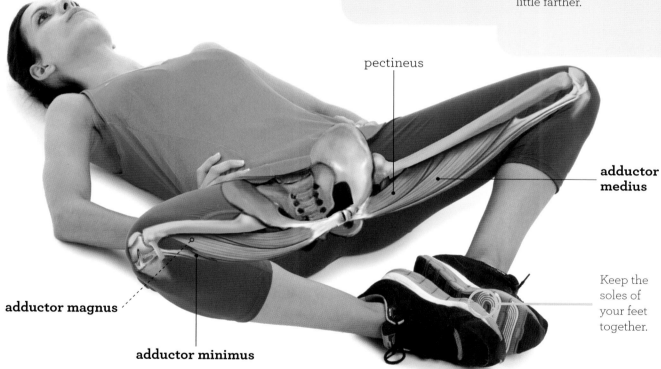

pectineus

adductor medius

adductor magnus

adductor minimus

Keep the soles of your feet together.

Starting Position

CAUTION
Don't apply too much pressure to your thighs, especially if you don't do this stretch regularly.

INDICATION
For athletes who run cross-country events, hurdle events, or obstacle events, plus for players of team sports involving running or racquets.

Adduction with Side Support

START

Choose a support about 16 inches (40 cm) high, such as a bench, chair, or stool, and rest your forearm on it. The leg on the support side should be straight, and the outside of your foot should touch the floor. The other leg must cross over the first one, with your knee bent around 90° and the sole of your foot flat on the floor.

TECHNIQUE

Relax the position so that your hip lowers and increases its degree of adduction. Use the support from your hands and your crossed leg to keep the movement slow and controlled through its entire range.

Support your forearm solidly on the bench.

gluteus minimus

gluteus medius

tensor fasciae latae

gluteus maximus

Starting Position

LEVEL	SETS	DURATION
BEGINNER	2	20 sec
INTERMEDIATE	3	20 sec
ADVANCED	3	30 sec

CAUTION

Use a support that assures stability and has a sufficiently broad, stationary base that will not move with the motion of your body.

INDICATION

For runners who do hurdle and obstacle events, and, especially, for long-distance runners and for athletes who play racquet and other sports that involve lateral movement.

Standing Rear Crossover

gluteus minimus

gluteus medius

gluteus maximus

tensor fasciae latae

Point your hands toward the crossed foot.

Starting Position

START
Cross one foot behind the other, bend your upper body slightly forward, and let your arms hang down relaxed. The knee of the forward leg will need to be slightly bent, but the other one is totally straight.

TECHNIQUE
Keep bending your upper body forward and toward the side where your crossed foot is located. Point your hands toward the crossed foot and try to come as close to it as possible. You can slightly increase the bend in the forward knee, but keep the other one totally straight during the whole exercise.

LEVEL	SETS	DURATION
BEGINNER	2	20 sec
INTERMEDIATE	2	20 sec
ADVANCED	3	30 sec

CAUTION
Start from a stable position, and remember that your center of gravity will change with the lean of your upper body.

INDICATION
For long-distance runners, because of the way these events involve the iliotibial band. Also for hurdlers and athletes who do racquet sports, judo, tae kwon do, volleyball, and any other sports that regularly involve lateral movement.

Standing Rear Crossover with Support

START
Rest both hands on a raised support, such as the back of a bench or anything else you have available for doing your exercises. Lean your upper body forward slightly and put one foot behind the other. The knee of the rear leg must be straight.

TECHNIQUE
Bend the knee of the forward leg to lower your center of gravity, and slide the rear foot on its outside surface so that one leg crosses behind the other. The rear leg must remain straight throughout.

gluteus minimus

gluteus medius

gluteus maximus

tensor fasciae latae

Slide the rear foot on its outer surface.

Starting Position

LEVEL	SETS	DURATION
BEGINNER	2	20 sec
INTERMEDIATE	3	20 sec
ADVANCED	3	30 sec

CAUTION
Use a sturdy support, because you will have to hold onto it to keep your balance.

INDICATION
For long-distance runners and those runners who participate in hurdle and obstacle events.

Knight's Position

Straighten your hip as much as possible.

psoas major

ileopsoas

sartorius

START
Place one knee on the floor with a bend of about 90°. The other leg is out front, also with the knee bent around 90° and the foot flat on the floor. You can place your hands on your hips, on the thigh of the forward leg, or hold onto something, if you find it difficult to keep your balance. In any case, keep your upper body perpendicular to the floor.

TECHNIQUE
Move your body forward, which will produce a straightening in the hip corresponding to the trailing leg and increase the amount of bend in the forward knee. You will feel tension in the front of your pelvis as your body moves forward.

Starting Position

LEVEL	SETS	DURATION
BEGINNER	2	20 sec
INTERMEDIATE	3	25 sec
ADVANCED	3	30 sec

CAUTION
Be sure to maintain a stable position and use a mat whenever possible.

INDICATION
For sprinters, hurdlers, and obstacle course runners, as well as for people who experience discomfort in their lower spines due either to excessive curvature or to anteversion of the pelvis.

Low Stride

START

Place one foot as far ahead of the other as possible while still allowing you to keep your balance reliably. The forward foot must rest flat on the floor, and the front part of the other foot touches the floor. Keep your back straight and perpendicular to the floor and your arms relaxed at your sides (or else on the thigh of the forward leg).

TECHNIQUE

Bend the knee of the forward leg, so that your center of gravity lowers and moves forward. The hip of the rear leg should reach the maximum extension possible. You can rest both hands on the thigh of the forward leg to improve your stability and control over the movement.

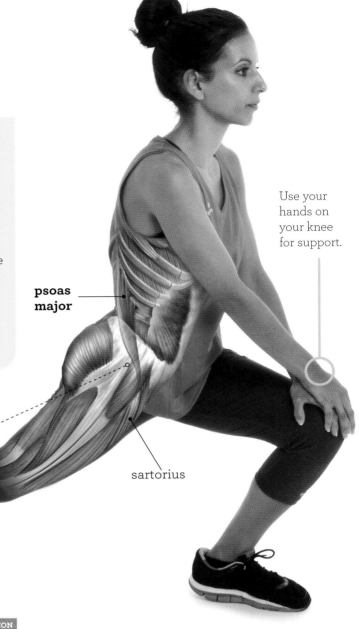

Use your hands on your knee for support.

psoas major

ileopsoas

sartorius

Starting Position

LEVEL	SETS	DURATION
BEGINNER	2	20 sec
INTERMEDIATE	3	25 sec
ADVANCED	3	30 sec

CAUTION

Check that your feet are the right distance apart by doing the first part of the movement before doing the whole exercise.

INDICATION

For sprinters, hurdlers, obstacle course runners, and trail runners, plus people who experience discomfort in their lower backs due to anterversion of the pelvis.

Supine with Crossed Leg

Starting Position

START

Lie down on your back and rest one foot on the floor. The corresponding knee should be bent to about 75°. The other leg should be crossed over the first one and relaxed in such a way that the knee is bent and the foot hangs down. You can place your hands behind your head or keep your arms relaxed at your sides.

TECHNIQUE

Use your upper leg to pull the other one toward the midline of your body, producing adduction in the bent hip. Keep the lower part of your back on the floor while pulling, and you will feel the tension in the gluteus. Hold the position for a few seconds.

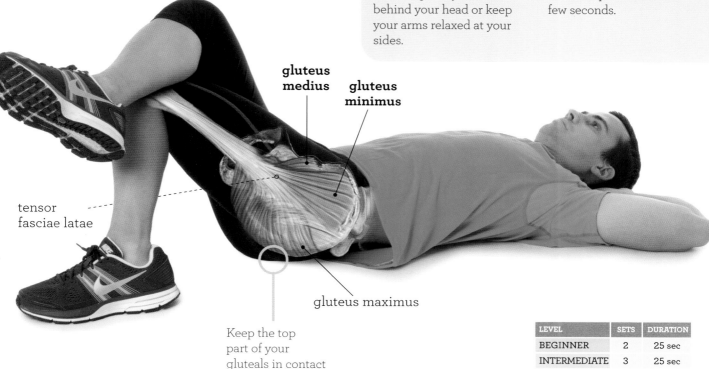

gluteus medius

gluteus minimus

tensor fasciae latae

gluteus maximus

Keep the top part of your gluteals in contact with the floor.

LEVEL	SETS	DURATION
BEGINNER	2	25 sec
INTERMEDIATE	3	25 sec
ADVANCED	3	30 sec

CAUTION

Keep the lower part of your back in contact with the floor or very close to it, even though this reduces the range of the exercise, so you don't detract from the technique of the exercise or its effectiveness.

INDICATION

For sprinters, hurdlers, and trail runners, because of the special power they need to apply to create momentum. These athletes make particular use of the gluteals in dealing with uneven ground, jumping, and accelerating as they run.

Crossed Leg Pull

START

Sit down with one leg straight out in front of you. Cross the other leg over it, with your knee bent, so that the outside of your foot touches the floor. Place one hand on the bent knee and use the other hand for support.

TECHNIQUE

Pull on the knee you are holding, so that it rises and your hip experiences adduction and flexion. The foot corresponding to the crossed leg should be flat on the floor. Don't let the pulling rotate your upper body or lift your gluteal from the floor.

gluteus medius

gluteus minimus

gluteus maximus

tensor fasciae latae

Keep the gluteal being stretched in contact with the floor.

Starting Position

LEVEL	SETS	DURATION
BEGINNER	2	25 sec
INTERMEDIATE	3	25 sec
ADVANCED	3	30 sec

CAUTION

Make sure that the crossed foot is well supported and that you have a firm, sure grip on your knee.

INDICATION

For sprinters and runners in hurdle and obstacle events, as well as in cross-country and trail running.

Supine Knee Pull

Starting Position

START

Lie down on your back and bend the hip and knee of one leg as you keep the other leg straight and relaxed. Place the opposite hand on the bent knee and hold it firmly. Your free hand can be relaxed, with your arm nearly in line with your shoulders.

TECHNIQUE

Pull on the bent knee, so that the leg crosses over your body and your hip is bent as much as possible in adduction. Hold the pull at the point of maximum stretch to optimize the results of the exercise.

tensor fasciae latae

Rest your head on the floor to spare your neck from sustained tension.

gluteus medius　　**gluteus minimus**　　gluteus maximus

LEVEL	SETS	DURATION
BEGINNER	2	25 sec
INTERMEDIATE	3	25 sec
ADVANCED	3	30 sec

CAUTION

This exercise involves no special risk, so all you have to do to get good results is to use the proper technique.

INDICATION

For sprinters, hurdlers, obstacle course and cross-country runners, as well as for people who experience discomfort in the area of the gluteals.

Knee Pull to Chest

START

Sit with one leg stretched out straight and the other one crossed over it with a slight bend in the knee. Grasp that knee with both hands and keep your back straight. Both feet touch the floor with the heels.

TECHNIQUE

Pull the knee you are holding toward your chest, producing maximum flexion in the hip and knee, and pulling your foot off the floor, but keeping it crossed over the relaxed leg. When you feel the tension in your gluteals, stop the movement and hold the traction for a few seconds.

Pull on your knee with both hands.

gluteus maximus

gluteus medius

gluteus minimus

Starting Position

LEVEL	SETS	DURATION
BEGINNER	2	25 sec
INTERMEDIATE	2	30 sec
ADVANCED	3	30 sec

CAUTION

You will need to use both hands to produce the necessary pull, so make sure you start from a stable position.

INDICATION

For cross-country runners and, especially, for trail runners. Also for hurdlers, sprinters, and triathletes and for people who experience discomfort in the gluteal area.

The Chair

Keep your chest straight and relaxed.

pyramidal

mellus muscles

obturatorius internus

Starting Position

START
Cross one foot over the top of the support knee, which is kept straight. The ankle rests on the knee, and your back is straight and perpendicular to the floor, with your arms relaxed at your sides.

TECHNIQUE
Gradually bend your support knee and hip until you are nearly in a sitting position. You will need to lean your upper body forward to keep your balance, and your hands will slide as far as your knee and the inside of your foot. This will add to the stability of the position.

LEVEL	SETS	DURATION
BEGINNER	2	15 sec
INTERMEDIATE	2	20 sec
ADVANCED	2	25 sec

CAUTION
This exercise requires good balance and the use of considerable force in the support leg, so you will have to do the movement in a gradual and controlled manner.

INDICATION
For people who experience discomfort in the gluteal area, for sprinters, and for trail runners.

Supine Pull to Chest

START

Lie down on your back with your arms relaxed at your sides. Bend one knee to about 90°, so that the sole of your foot is flat on the floor. Cross the other leg over the first one and let it rest there.

TECHNIQUE

Hold the thigh of the lower leg with both hands and pull it toward your chest. This will produce flexion in both hips, and your knees will move closer to your chest. After stopping the movement, hold the traction for the appropriate time based on your level.

Starting Position

LEVEL	SETS	DURATION
BEGINNER	2	15 sec
INTERMEDIATE	2	20 sec
ADVANCED	2	25 sec

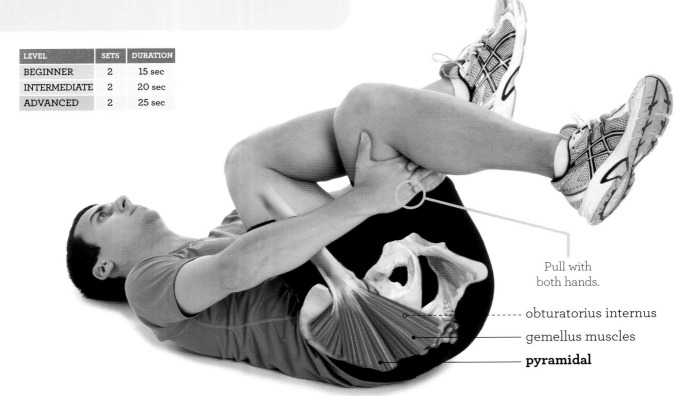

Pull with both hands.

obturatorius internus

gemellus muscles

pyramidal

CAUTION

Rest your head against the floor to avoid unnecessary tension in your neck muscles.

INDICATION

For all runners. Especially for sprinters, trail runners, hurdlers, and jumpers, and for people who experience discomfort in the gluteal area.

The Chair with Support

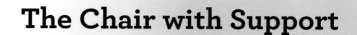

pyramidal
gemellus
muscles

obturatorius
internus

Keep your
ankle in place
on its original
support point.

Starting
Position

START
Sit on a bench or stool. Place one foot on the floor, so that your knee is bent at 90°, and place the outside of the opposite ankle on top of it. One hand should rest on the ankle and the other on the knee of the same leg.

TECHNIQUE
The hand resting on the knee presses down on it and moves it downward while you hold your ankle in place. When you reach the optimum stretching point, stop the movement and maintain light pressure to prolong the tension in the muscles being stretched.

LEVEL	SETS	DURATION
BEGINNER	2	15 sec
INTERMEDIATE	2	20 sec
ADVANCED	2	25 sec

CAUTION
Avoid applying excessive pressure to your knee, because the muscles to be stretched are relatively small.

INDICATION
For people who experience discomfort in the gluteal area, for sprinters, and for cross-country runners.

LOWER LIMB STRETCHES

QUADRICEPS FEMORIS

This is made up of four muscles that act as a functional unit and are located in the front of the thigh. Their main function is straightening the knee, so they keep it from bending upon contact and during practically the entire support phase in running. They contribute particularly to absorbing the impact when the foot makes initial contact with the ground. The following are the muscles that make up the quadriceps.

Rectus femoris: This muscle originates at the anteroinferior iliac spine and shares an insertion, with slight variations, with the crural muscle and the vastus externus and internus muscles, at the kneecap, via the quadriceps tendon, and at the tibial tuberosity, via the patellar ligament.

Crural muscle: This originates at the anterior and lateral faces of the diaphysis of the femur.

Vastus externus: This originates at the linea aspera femoris, the trochanter magnus, and the gluteal tuberosity of the femur.

Vastus internus: This originates along the linea aspera femoris.

ISCHIOTIBIALS

These are three muscles that share functions of straightening the hip and bending the knee, so their action is crucial to every contact phase in running, especially in the initial part, because they help control hip flexion. They also play a role in the forward thrust at the end of the support phase and the first phase of the leg swing.

Biceps femoris: Its long head originates at the ischium and the sacrotuberous ligament, and the short head originates at the linea aspera femoris and the lateral condyle of the femur. Its insertion is at the head and the lateral face of the fibula and at the lateral tibial condyle.

Semitendinosus: This muscle originates at the ischium, and it inserts at the proximal diaphysis of the tibia and the goosefoot.

Semimembranosus: This muscle originates at the ischium, and it inserts at the medial condyle of the tibia.

GASTROCNEMIUS (CALF MUSCLE)

This muscle's main function is plantar flexion of the ankle, so it plays an active role in the middle and final parts of the support phase, largely contributing to the forward thrust for the takeoff and the aerial, or flight, phase. It originates at the medial and lateral condyles of the femur, and it inserts at the heel bone by means of the Achilles tendon.

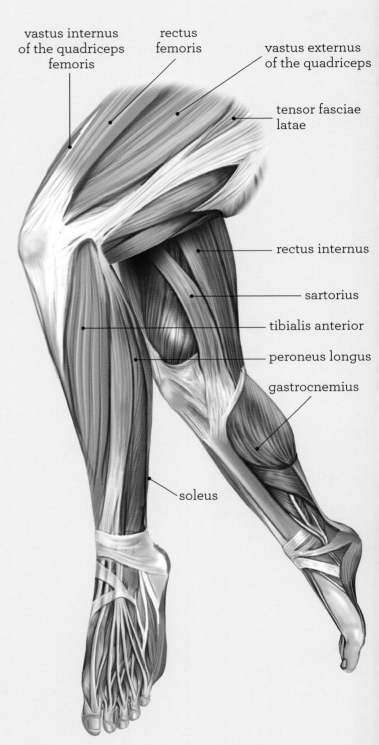

vastus internus of the quadriceps femoris

rectus femoris

vastus externus of the quadriceps

tensor fasciae latae

rectus internus

sartorius

tibialis anterior

peroneus longus

gastrocnemius

soleus

SOLEUS

This muscle shares its main function with the gastrocnemius, so its contribution to running is the same or very similar. It originates at the head and the third proximal of the diaphysis of the fibula and at the soleal line of the tibia. Its insertion at the heel bone via the Achilles tendon is shared with that of the gastrocnemius.

TIBIALIS ANTERIOR

The main function of this muscle is dorsal flexion of the ankle, so it contributes to shock absorption and ankle stability in the initial contact phase with the ground. It also is responsible for keeping the front of the foot raised during the entire leg swing, which facilitates elevation of the limb and prevents stumbling. It originates at the tibial condyle, the two proximal thirds of the tibial diaphysis, and the interosseous membrane, and it inserts at the first cuneiform and the first metatarsal.

PERONEUS

This is made up of three muscles with different functions, which will be described below, plus a fourth one, which will not be described, because it is found in only a few individuals and it brings no functional advantage to those who have it.

Peroneus longus: Its main function is outward rotation of the ankle, so it contributes to ankle stability, especially during the support phase. It originates at the head and the two proximal thirds of the fibular diaphysis, and it inserts at the first cuneiform and the first metatarsal.

Peroneus brevis: Its main function is also the outward rotation of the ankle, and it contributes to ankle stability while running. It originates at the two distal thirds of the fibular diaphysis, and it inserts at the fifth metatarsal.

Peroneus tertius: This is the smallest of the three peroneus muscles, and its main function is dorsal flexion of the ankle, so it acts in conjunction with the tibialis anterior in the leg-swing phase and at the moment of contact with the ground, by keeping the front of the foot raised. It originates in the third distal of the fibula, and it inserts at the fifth metatarsal.

PLANTAR FASCIA

This is not a muscle, but rather a tough, triangular membrane of connective tissue located on the sole of the foot. It is responsible for holding up the arch of the foot, and thus much of the structural integrity of the foot under the tensions of running. In addition, it is the anchor point for some of the foot muscles.

Its insertions are at the lower face of the heel bone and the first phalanges.

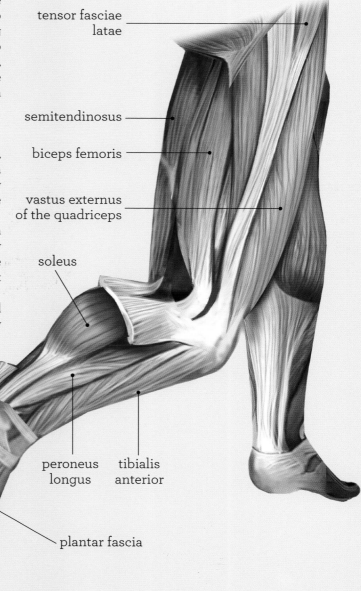

tensor fasciae latae

semitendinosus

biceps femoris

vastus externus of the quadriceps

soleus

peroneus longus

tibialis anterior

plantar fascia

Rearward Lean

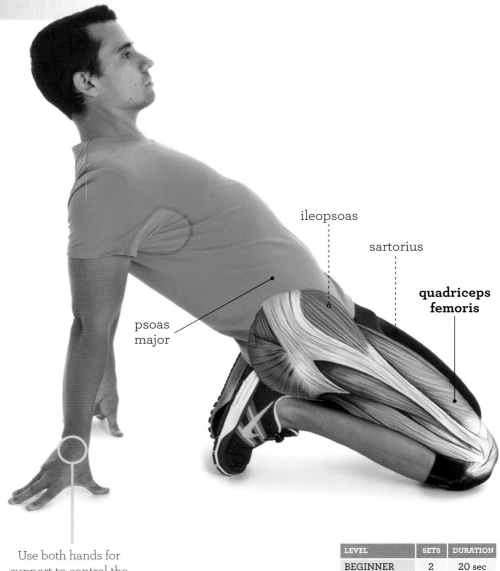

ileopsoas

sartorius

quadriceps femoris

psoas major

START

Kneel down with your legs together and your feet resting on tiptoes. Keep your back straight and perpendicular to the floor, with your arms relaxed at your sides.

TECHNIQUE

Lean rearward as far as you can without moving from your support points. Bend your knees to the maximum, so that the back of your thighs touch your calves. Your hips will straighten, and you will have to use your hands for support behind you. If you have not reached the best stretching point, bend your elbows and lower your body until you feel adequate tension in the front of your thighs.

Use both hands for support to control the lean of your upper body.

Starting Position

LEVEL	SETS	DURATION
BEGINNER	2	20 sec
INTERMEDIATE	3	25 sec
ADVANCED	3	35 sec

CAUTION

You must lean your body slowly and safely, and you will have to support yourself with your hands as soon as you feel the need.

INDICATION

For all runners, because of the work of the quadriceps in the contact phase and the initial and middle parts of the support phase. Especially for jumpers, hurdlers, and anyone who takes part in obstacle events, because of the particular abruptness of the contact and initial support phases after going over an obstacle.

Assisted Knee and Hip Extension

Starting Position

LEVEL	SETS	DURATION
BEGINNER	2	20 sec
INTERMEDIATE	3	25 sec
ADVANCED	3	35 sec

START
Lie face down with one knee straight and the other one bent to about 90°. Your assistant, or coach, should hold the latter leg with one hand above the knee and the other on the ankle or foot. Relax your upper body and fold your arms, so you can rest your head on them.

TECHNIQUE
The assistant needs to pull on your foot or ankle to create the maximum possible bend in your knee. If you do not feel enough tension in the front of your thigh, your assistant will have to lift your knee slightly off the floor, emphasizing the hip extension without reducing the bend in your knee.

Your knee will have to come off the floor.

quadriceps femoris   sartorius   ileopsoas   psoas major

CAUTION
Remember that your assistant has no way of knowing how much stretch you are feeling in your quadriceps, so you need to communicate throughout this exercise.

INDICATION
For all runners, whether short or long distance. Especially for athletes in events involving obstacles or cross-country, where runners must negotiate long, steep grades, whether uphill or downhill.

Unilateral on Side

START

Lie down on one side on the floor. The leg closer to the floor should be aligned with your upper body, and the arm on this side should form a right angle to your body so it can serve as support during the exercise. Use your free hand to hold the ankle of the upper leg, so that your knee is bent.

TECHNIQUE

Pull your ankle rearward as if you were trying to touch your gluteal with the sole of your foot, so that your knee is totally bent and your hip is straight. You will feel the tension in the front of your thigh, which you can increase gradually by straightening your hip further.

Starting Position

Pull on your ankle to create maximum hip extension.

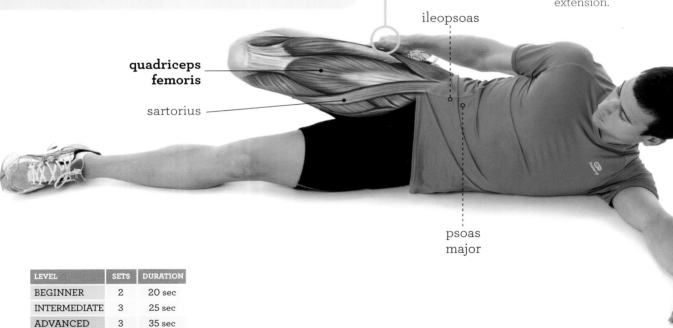

ileopsoas

quadriceps femoris

sartorius

psoas major

LEVEL	SETS	DURATION
BEGINNER	2	20 sec
INTERMEDIATE	3	25 sec
ADVANCED	3	35 sec

CAUTION

This exercise involves no risk, but you need to find a starting position stable enough so that you don't have to reposition yourself partway through.

INDICATION

For all runners. Especially for athletes who need to jump over obstacles, cover long distances, or negotiate steep slopes uphill and downhill.

Flamingo Position

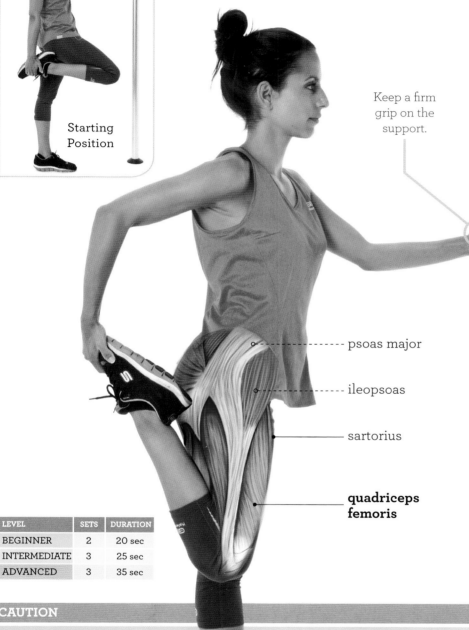

Starting Position

Keep a firm grip on the support.

- - - - - - - - - psoas major

- - - - - ileopsoas

—— sartorius

quadriceps femoris

LEVEL	SETS	DURATION
BEGINNER	2	20 sec
INTERMEDIATE	3	25 sec
ADVANCED	3	35 sec

START
Stand facing an object you can use for support and hold onto it with one hand. Bend the opposite knee and hip and hold your ankle or instep with your free hand, so that your leg is tucked up. Your upper body must remain perpendicular to the floor so that your body weight falls on the support leg. The hand holding the support can keep you steady.

TECHNIQUE
Pull your instep rearward as if you were trying to touch your heel to your gluteal, producing maximum hip extension. The knee of the leg you are holding must remain bent throughout in order to maximize the effect of the stretch.

CAUTION
Use a solid support you can hold onto with your free hand to keep you steady while you do the stretch.

INDICATION
For all runners. Especially for those who have to jump over obstacles, cover long distances, or negotiate steep slopes, whether uphill or downhill.

Knight's Position with Pull

START

Take a position on one knee and one foot, with one leg ahead of the other and both knees and one hip bent to about 90°, so that you are in a position similar to the one that a medieval knight would take during the knighting ceremony. Rotate your upper body so that your trailing foot and the corresponding hand are close together. The other hand rests on the forward knee.

TECHNIQUE

Hold the ankle of the rear foot with the corresponding hand and pull on it, bending your knee as far as possible, while you lean your upper body forward slightly to increase the hip extension.

LEVEL	SETS	DURATION
BEGINNER	2	20 sec
INTERMEDIATE	3	25 sec
ADVANCED	3	35 sec

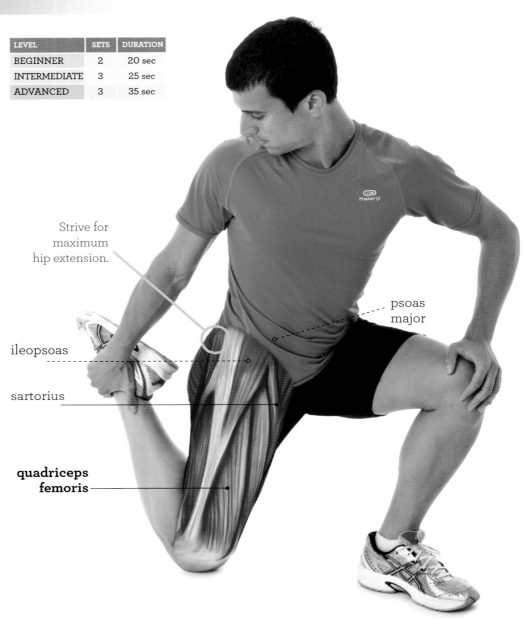

Strive for maximum hip extension.

psoas major

ileopsoas

sartorius

quadriceps femoris

Starting Position

CAUTION

Be sure to start from a balanced position, and do the exercise on a padded or smooth surface, so you don't injure the support knee.

INDICATION

For people who experience tension in the front of the thigh who participate in team sports that involve running, especially if those sports require bursts of speed, sudden braking, or abrupt changes of direction, and for long-distance and cross-country runners.

Hip Flex

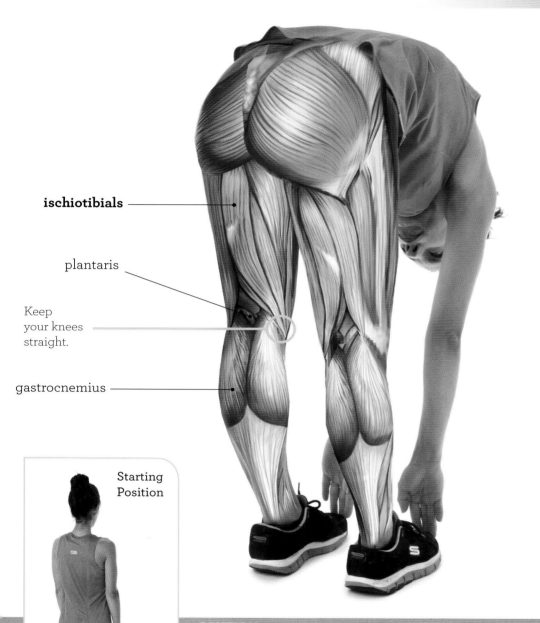

ischiotibials

plantaris

Keep your knees straight.

gastrocnemius

Starting Position

START
Stand with your back straight and your arms relaxed, and look straight ahead. Keep your feet a short distance apart to assure stability while you do this exercise.

TECHNIQUE
Lean your upper body forward by bending at the hips, and try to touch your toes with your fingers. It's essential to keep your knees straight during the entire process, because doing otherwise would completely cancel out the stretch of the ischiotibial muscles.

LEVEL	SETS	DURATION
BEGINNER	2	25 sec
INTERMEDIATE	3	30 sec
ADVANCED	3	35 sec

CAUTION
Keep your feet slightly apart at all times so you can keep your balance while doing this exercise. Distance has no influence on the effectiveness of the stretch.

INDICATION
For people who experience tension in the back of the thigh or discomfort in the lower back due to retroversion of the pelvis. Also for sprinters, cross-country runners, and athletes who play team sports involving running, especially if sudden bursts of speed are called for.

Foot Pull with Towel

START
Lie down on your back and bend your knees to about 90°. Rest one foot on the floor and raise the other one. Place a towel, an article of clothing, or something similar under the sole of your foot and hold each end. The towel should pull on your heel as you do this stretch.

TECHNIQUE
Straighten the knee of the raised leg as you pull on the towel. This will bring your foot nearly perpendicular to your head. Keep the lower part of your back in contact with the floor, and you will feel the tension from the stretch in the back of your thigh and knee.

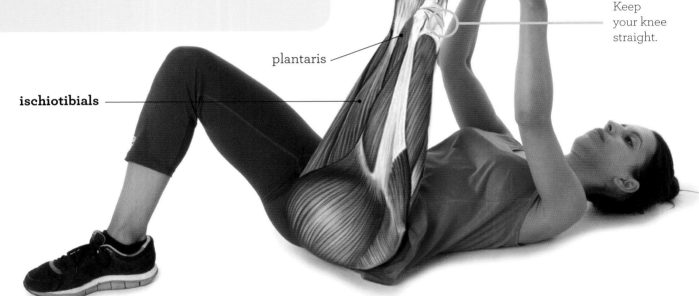

Keep your knee straight.

plantaris

ischiotibials

Starting Position

LEVEL	SETS	DURATION
BEGINNER	2	20 sec
INTERMEDIATE	3	25 sec
ADVANCED	3	35 sec

CAUTION
Do the stretch slowly and gradually, because the ischiotibial muscles are particularly sensitive to stretching.

INDICATION
For sprinters, cross-country runners, and people who experience tension in the back of the leg or tightness in the ischiotibial muscles. The latter is very common in soccer players and athletes who play other team sports involving running.

Hip Flex with Support

Starting Position

LEVEL	SETS	DURATION
BEGINNER	2	20 sec
INTERMEDIATE	3	25 sec
ADVANCED	2	35 sec

START
Rest one foot on a raised support, keeping the hip of the raised leg bent and your knee straight. Place the palms of your hands on your thigh and knee, and keep your back straight and perpendicular to the floor. The knee of the support leg should be straight, although there can be a slight bend if you are using a very low support or if you have lots of flexibility in your ischiotibial muscles.

TECHNIQUE
Slide your hands forward and try to touch your toes as you lean your upper body forward, which will increase the bend in your hip. Avoid bending the knee of the leg being stretched, and hold the position for a few seconds once you reach the point of optimum stretch.

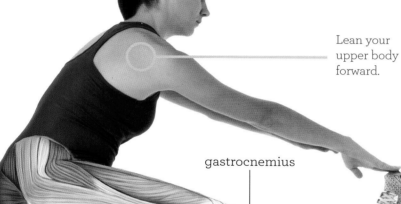

Lean your upper body forward.

gastrocnemius

ischiotibials

plantaris

CAUTION
Use a solid support and do the exercise slowly and gradually.

INDICATION
For sprinters and people who experience retroversion of the pelvis as a result of tightness or tension in the ischiotibial muscles.

Assisted Supine Hip Flex

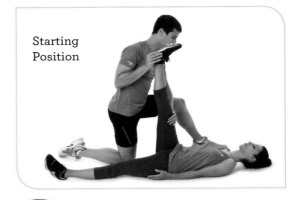

LEVEL	SETS	DURATION
BEGINNER	2	20 sec
INTERMEDIATE	3	25 sec
ADVANCED	3	35 sec

START

Lie down on your back with one leg in line with your upper body and the other one forming a 90° angle with your upper body. Your assistant, or coach, should take a place next to you and should hold your raised leg by the heel and the front of the knee.

TECHNIQUE

Your assistant pushes the leg toward you. The thrust comes from the hand holding your heel, while the other hand keeps the knee straight. This will emphasize the hip bend and produce the stretch in the ischiotibial muscles.

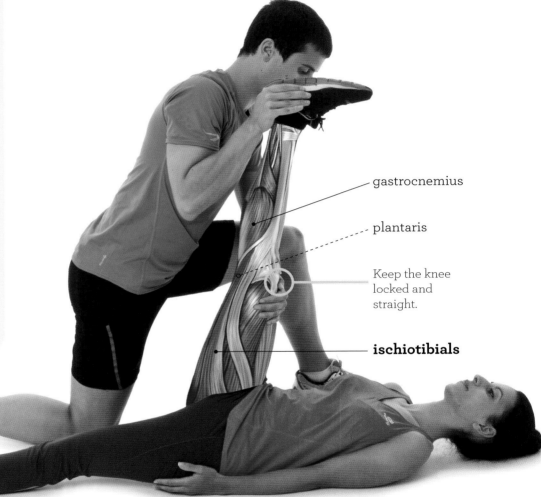

gastrocnemius

plantaris

Keep the knee locked and straight.

ischiotibials

CAUTION

Make sure that the person helping you executes the movement slowly, and that both of you communicate continually, so the movement can be stopped at the right instant. This will optimize the results and minimize the risk of injury.

INDICATION

For sprinters, for athletes who play sports with quick starts or sudden changes of speed, and for people who experience tension in the back of their thighs or tightness in their ischiotibial muscles.

Assisted Standing Hip Flex

Starting Position

Bend your hip as far as you can while maintaining safety.

plantaris

ischiotibials

START
Stand in front of your assistant. Raise one leg so he or she can hold your foot by the heel and the instep. The support leg must be straight at the knee, and your hands can rest on the raised leg to improve your stability and comfort during this exercise.

TECHNIQUE
Your assistant moves the raised leg upward, thereby emphasizing the bend in your hip, while your knee remains straight. Your upper body should remain perpendicular to the floor or lean forward slightly if the stretch is not adequate.

LEVEL	SETS	DURATION
BEGINNER	2	20 sec
INTERMEDIATE	3	25 sec
ADVANCED	3	35 sec

CAUTION
To minimize the risk of injury, make sure your assistant moves slowly and that both of you communicate throughout.

INDICATION
For sprinters, for participants in sports involving running, and, especially, for athletes whose sports require fast starts or sudden changes of cadence. Also for people who experience retroversion of the pelvis, because of tightness in the ischiotibials.

Squat with One Foot Forward

START

Stand with one foot ahead of the other, both knees straight, your upper body perpendicular to the floor, and your hands resting on the rearmost thigh.

TECHNIQUE

Shift your weight to the rear leg at the same time that you bend your knee and use the support from your hands to control the movement. Keep the forward leg straight at the knee, and slide the foot forward as you lower your body's center of gravity.

LEVEL	SETS	DURATION
BEGINNER	2	20 sec
INTERMEDIATE	3	25 sec
ADVANCED	3	35 sec

Starting Position

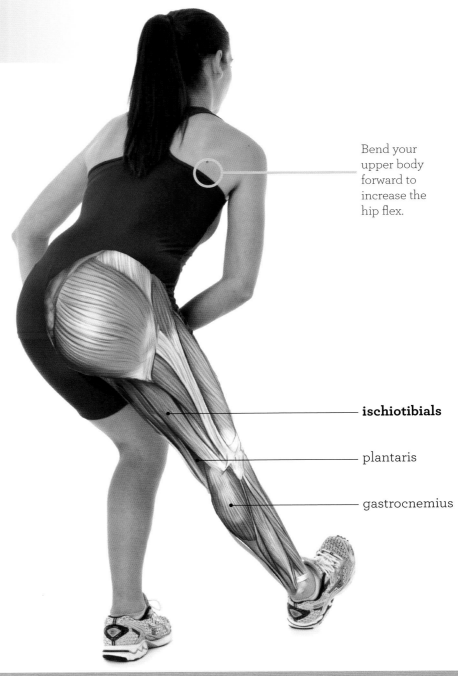

Bend your upper body forward to increase the hip flex.

ischiotibials

plantaris

gastrocnemius

CAUTION

This exercise necessarily involves significant instability, so you need to begin from a balanced position and move slowly and deliberately.

INDICATION

For athletes who play sports that involve running, especially if those sports require quick starts or sudden accelerations. Also for people who experience tension in the back of the thigh or tightness in the ischiotibial muscles.

V-seat

Starting Position

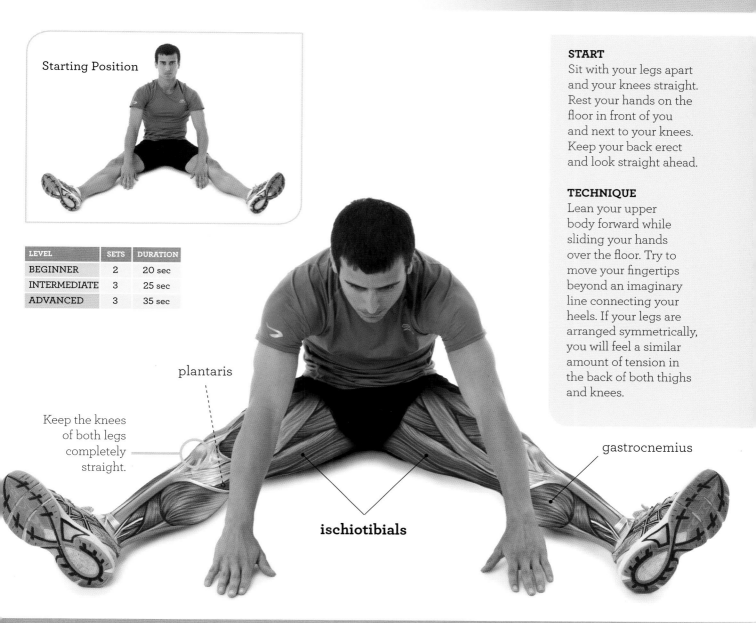

LEVEL	SETS	DURATION
BEGINNER	2	20 sec
INTERMEDIATE	3	25 sec
ADVANCED	3	35 sec

plantaris

Keep the knees
of both legs
completely
straight.

gastrocnemius

ischiotibials

START
Sit with your legs apart
and your knees straight.
Rest your hands on the
floor in front of you
and next to your knees.
Keep your back erect
and look straight ahead.

TECHNIQUE
Lean your upper
body forward while
sliding your hands
over the floor. Try to
move your fingertips
beyond an imaginary
line connecting your
heels. If your legs are
arranged symmetrically,
you will feel a similar
amount of tension in
the back of both thighs
and knees.

CAUTION
Lean your upper body
forward without bending
your spine excessively,
because that would not
contribute to the stretch and
you would risk injury.

INDICATION
For sprinters and for athletes
who play sports involving
running with sudden starts
and accelerations. Also for
people who experience
tightness or tension in the
ischiotibial muscles, as well
as for those who suffer from
retroversion of the pelvis.

Seated Unilateral Hip Flex

START

Sit with your legs spread apart. One of your knees should be straight, and the other one should be bent with the sole of your foot flat on the floor. Your upper body should be perpendicular to the floor and slightly rotated, so that you are facing your straight leg. Place your arms next to your body with your elbows bent, as if you were a boxer.

TECHNIQUE

Lean your upper body toward your straight leg, as if you were trying to bring your chest close to it by bending at the hip, while you keep your knee totally straight to force the stretch in your ischiotibial muscles.

Starting Position

LEVEL	SETS	DURATION
BEGINNER	2	20 sec
INTERMEDIATE	3	25 sec
ADVANCED	3	35 sec

Keep your knee locked and straight.

ischiotibials

plantaris

gastrocnemius

CAUTION

Lean forward but do not bend your spine excessively, because doing so does not contribute to the stretch, and it could entail a risk of injury.

INDICATION

For sprinters and for athletes who perform sudden starts and accelerations. Also for people with retroversion of the pelvis due to tightness or tension in the ischiotibial muscles.

Supine Dorsal Ankle Flex

soleus

gastrocnemius

Keep your
knee straight.

plantaris

START
Lie down on your back
and rest one foot on the
floor, so that your knee is
bent to about 90°. Raise
the other leg, keep your
knee straight, and hold
onto it with both hands.
The raised leg must
remain perpendicular to
the floor, with your ankle
in plantar flexion.

TECHNIQUE
Lock the knee of the
raised leg straight and
put your ankle into dorsal
flexion. As you emphasize
the dorsal flexion, the
tension in the back of
your leg will increase. This
will be an unmistakable
sign of the stretch in your
gastrocnemius muscles.

LEVEL	SETS	DURATION
BEGINNER	3	20 sec
INTERMEDIATE	3	25 sec
ADVANCED	3	30 sec

Starting
Position

CAUTION
Keep your head on the
floor to avoid unnecessary
tension in your cervical
vertebrae.

INDICATION
For all runners, especially those
who cover long distances, such as
marathoners. Also recommended
for athletes who play sports in which
events or games go on for a long
time and in which there are rest
periods or slack time that they can
use to reduce accumulated tension.

Static Stretches After Running / **107**

Push-up Position

Starting
Position

START

Take a position similar
to the top phase of a
push-up, in which you
are supported on the
palms of your hands and
the front of your feet,
without your upper body,
thighs, or knees touching
the floor. Then draw up
one leg by bending at the
hip and knee. The sole
of the foot that stays
in place should be
perpendicular to the
floor.

TECHNIQUE

Place the ankle of your
anchor foot into dorsal
flexion, keeping your knee
straight and bringing
your body a short distance
to the rear. You will feel
the tension in the back
of your leg, which you
should maintain for a few
seconds.

Keep
your back
straight.

plantaris

gastrocnemius

soleus

LEVEL	SETS	DURATION
BEGINNER	3	20 sec
INTERMEDIATE	3	25 sec
ADVANCED	3	30 sec

CAUTION

Keep your back straight
and your abdominal
muscles tight. This will
help avoid sagging toward
the floor and the consequent
risk of injury to the lower
spine.

INDICATION

For all runners, especially
long-distance runners
or those who suffer from
cramps or strain in the
calf area.

Toe Pull

Starting Position

START

Assume the knight's position—that is, get down on one knee with the opposite foot out in front of you. In this position, the hip and knee of the forward leg will be bent at 90°, and the hip of the rear leg will be straight. Starting from this position, slightly straighten the knee of the forward leg, lean your upper body forward, and hold the toe of your shoe with one hand.

TECHNIQUE

Straighten the knee of the forward leg all the way by moving your body rearward a short distance at the same time as you pull on the toe of your shoe to keep your ankle in dorsal flexion. You will feel the tension in your calf and, perhaps, in the back of your knee and thigh, even though the stretch of the latter area is secondary in this exercise.

LEVEL	SETS	DURATION
BEGINNER	3	20 sec
INTERMEDIATE	3	25 sec
ADVANCED	3	30 sec

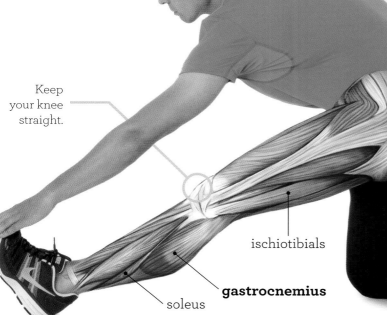

Keep your knee straight.

ischiotibials

gastrocnemius

soleus

CAUTION

Focus on the stretch in the back of the calf, to give higher priority to the stretch in the muscles responsible for plantar ankle flexion over the stretch in the ischiotibials.

INDICATION

For athletes who experience tension in the calf, for long-distance runners, or for people who play sports that involve running and extended periods of play, such as soccer and tennis.

Solid Support Push

START

Stand in front of a vertical support you can hold onto, such as a grate, a lamp post, a tree, or some other item you can find during your workout. Hold onto this support with both hands, at a distance that allows you to keep both elbows straight. Place one foot perpendicular to your upper body and the other one on tiptoes a fair distance behind the first one.

TECHNIQUE

Bend the knee of the forward leg to about 130° and lean your upper body forward, forming a straight line with the rearmost leg. The foot of the rear support, including the entire heel, should be flat on the floor, which will force the dorsal flexion of the ankle.

Starting Position

LEVEL	SETS	DURATION
BEGINNER	3	20 sec
INTERMEDIATE	3	25 sec
ADVANCED	3	30 sec

plantaris

gastrocnemius

soleus

CAUTION

Use a solid vertical support and hold onto it with both hands.

Place the heel of the rear foot on the floor.

INDICATION

For long-distance and cross-country runners, as well as for people who play sports involving running and lengthy play time.

Assisted Dorsal Ankle Flex

START

Lie down on your back and raise one leg to an angle of about 40° with the floor. Your assistant should stand near your ankles and hold the heel and the toes of the raised foot. Keep your knees straight and your head resting on the floor.

TECHNIQUE

Your assistant should push on your toes while holding your heel. This will produce dorsal flexion in your ankle. Keep your knee straight during this exercise to create the best possible stretch in the gastrocnemius and soleus muscles.

Keep your knee straight.

soleus

plantaris

gastrocnemius

Starting Position

LEVEL	SETS	DURATION
BEGINNER	3	20 sec
INTERMEDIATE	3	25 sec
ADVANCED	3	30 sec

CAUTION

Communicate constantly with your assistant to make sure that the stretch is done gradually and to minimize the risk of injury.

INDICATION

For long-distance runners, for people who experience tension in the calf muscles, and for athletes who participate in individual or team events involving running and prolonged play times.

Inclined Flex

START

Stand in front of a solid support no higher than your waist, such as the back of a bench, a trash can, or some other item you encounter on your regular training run. Hold onto the support with both hands and lean forward, keeping your upper and lower body in line. Keep your feet flat on the ground, from heel to toe.

TECHNIQUE

Slowly bend your elbows to increase the lean in your upper body and legs, keeping them in alignment as if they were a plank. The soles of your feet need to remain in total contact with the ground; your heels should not leave the ground at any time.

LEVEL	SETS	DURATION
BEGINNER	3	20 sec
INTERMEDIATE	3	25 sec
ADVANCED	3	30 sec

plantaris

gastrocnemius

soleus

Keep your heels on the ground.

Starting Position

CAUTION

Choose a support that's firmly anchored to the ground, and avoid doing this exercise on a surface where you could slip, such as on sand, gravel, or slick pavement.

INDICATION

For athletes whose sports involve running, especially if the playing times are long, as with tennis and soccer. Also for long-distance runners and swimmers.

"Set" Position

plantaris

gastrocnemius

soleus

Keep your knee straight and your heel on the ground.

START
Support yourself on your hands and feet. Your hands should be in line with one another, but one foot should be slightly ahead of the other. Keep your elbows straight and your knees bent a bit more than 90°, with your body in a position similar to the one used at the start of a sprint (when the starter gives the "set" command).

TECHNIQUE
Straighten your rear leg all the way and perform dorsal flexion with your ankle, so that your heel remains on the ground and you can feel the tension in your calf and the back of your knee and thigh.

Starting Position

LEVEL	SETS	DURATION
BEGINNER	3	20 sec
INTERMEDIATE	3	25 sec
ADVANCED	3	30 sec

CAUTION
Do this exercise on a nonslip surface and start from a balanced position.

INDICATION
For long-distance runners and swimmers and for other athletes in sports involving foot travel and long play times.

Bilateral Pull with Towel

START

Sit with your feet together in front of you. Your knees need to be slightly bent, so you can place a towel or other similar item on the front of the soles of your feet. Hold both ends of the towel firmly.

TECHNIQUE

Straighten your knees so that your calves remain in contact with the ground, and then pull on the ends of the towel to create dorsal flexion in your ankles. Even though this bend is not very pronounced, you will feel the tension in your calves and the stretch will be effective.

LEVEL	SETS	DURATION
BEGINNER	3	20 sec
INTERMEDIATE	3	25 sec
ADVANCED	3	30 sec

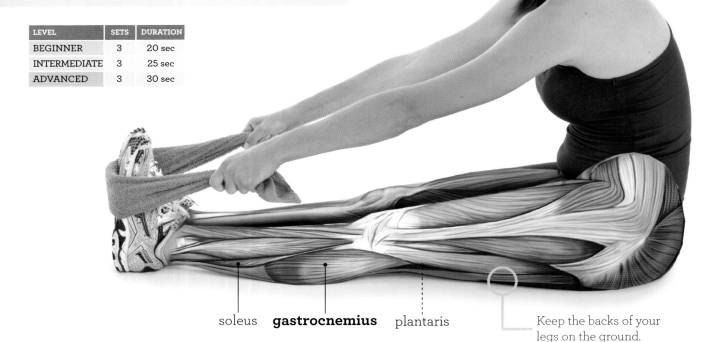

soleus **gastrocnemius** plantaris

Keep the backs of your legs on the ground.

Starting Position

CAUTION

The closer to the tip of your toes you place the towel, the more effective the exercise will be, but you have to find the right point so the towel does not slip off.

INDICATION

For long-distance runners, for people who experience tension in their calf muscles, for athletes who play sports involving running, and for athletes whose events go on for quite a while, such as in soccer, tennis, or basketball.

Squat Position

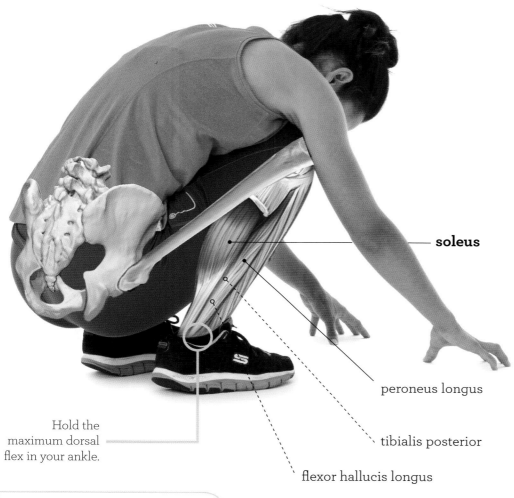

soleus

peroneus longus

tibialis posterior

flexor hallucis longus

Hold the maximum dorsal flex in your ankle.

START
Take a position with your feet shoulder-width apart and lean forward until you can touch the ground with the tips of your fingers. Your hands should be in front of your feet, and your knees must be bent to about 90°. At this point, your hips will be totally bent and your ankles will begin to experience a slight dorsal flexion.

TECHNIQUE
Squat down all the way and lean forward as much as possible, without picking your heels up from the ground. This way, you will emphasize the dorsal flexion of your ankles, which will produce enough tension in the soleus for the exercise to be effective.

LEVEL	SETS	DURATION
BEGINNER	2	20 sec
INTERMEDIATE	3	25 sec
ADVANCED	3	30 sec

Starting Position

CAUTION
Avoid raising your heels from the ground in order to force the upper body lean. This would not increase the intensity of the stretch and you could lose your balance.

INDICATION
For long-distance runners and for athletes who may experience tension or cramps in the calf area, such as swimmers, tennis players, soccer players, walkers, and cyclists.

"On Your Mark" Position

START

Support yourself on both hands, one foot, and one knee. Your hands should be ahead of your shoulders and slightly more than shoulder-width apart. Your support leg should be bent to about 75°, and your hip should be totally bent. Your trailing leg should be located slightly to the rear of the heel of the support foot, so your body will be in a position similar to the one used before the start of a sprint, when the official gives the command, "On your mark!"

TECHNIQUE

Slide your hands forward without modifying any of the other support points. This will move your upper body forward and your ankle will reach the point of maximum dorsal flexion. That will stretch the soleus.

tibialis posterior

soleus

peroneus longus

flexor hallucis longus

peroneus brevis

Reach the maximum point of dorsal flexion of the ankle.

Starting Positon

LEVEL	SETS	DURATION
BEGINNER	2	25 sec
INTERMEDIATE	3	25 sec
ADVANCED	3	30 sec

CAUTION

Avoid lifting the heel of your support foot, because doing so would contribute nothing to the exercise and would detract from the technique.

INDICATION

For long-distance runners, walkers, and other athletes who may place significant strain on the muscles used in plantar flexion of the ankle.

Shot-put Position

peroneus
longus

flexor hallucis
longus

soleus

tibialis
posterior

peroneus
brevis

Keep your
heel on the
ground.

Starting Position

START
Place one foot slightly
ahead of the other,
touching the ground
only with your toes. Bend
your knees and hips as if
you were about to squat,
while you lean your upper
body forward. Hold the
forward knee with both
hands.

TECHNIQUE
Lower the heel of the
rear foot until the entire
sole touches the ground.
If you feel tension in
your calf, the stretch is
occurring properly. If you
do not feel tension, you
need to start over, moving
the leg to be stretched
farther to the rear.

LEVEL	SETS	DURATION
BEGINNER	2	25 sec
INTERMEDIATE	3	25 sec
ADVANCED	3	30 sec

CAUTION

Make sure to start from a
stable position, so you can
do the exercise safely.

INDICATION

For long-distance runners,
walkers, and other athletes
who may strain the muscles
used in plantar flexion of
the ankle.

Seated Foot Pull

START
Sit on the ground with your legs together, your knees bent to about 100°, and your feet resting on your heels. Lean your upper body forward and place your fingers on your insteps.

TECHNIQUE
Hold your toes with your hands and pull back on them, trying to create the greatest possible dorsal flexion in your ankles. Your heels must remain in place on the original support points, but your knees may bend slightly more.

LEVEL	SETS	DURATION
BEGINNER	2	25 sec
INTERMEDIATE	3	25 sec
ADVANCED	3	30 sec

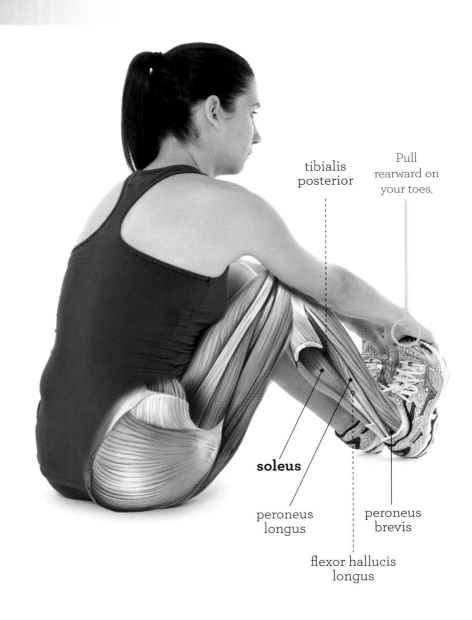

tibialis posterior

Pull rearward on your toes.

soleus

peroneus longus

peroneus brevis

flexor hallucis longus

Starting Position

CAUTION	INDICATION
Avoid straightening your knees when you do this exercise, so that it involves your soleus more completely.	For long-distance runners, walkers, and athletes who place special demands on the muscles used for plantar flexion of the ankle.

Knee Bends with Support

Starting Position

START
Stand in front of a support you can hold onto, whether vertical—such as a tree, a lamp post, or a pole—or horizontal—such as a table—as long as it is not too low. Hold onto the support with both hands and place one foot slightly ahead of the other. Bend your knees a little bit to lower your center of gravity slightly.

TECHNIQUE
Increase the degree of bend in your knees, further lowering your center of gravity, without changing the support points or lifting the heel of the rear leg from the ground. Your ankle will reach maximum dorsal flexion and the stretch will occur in your soleus. Hold the position for a few seconds.

LEVEL	SETS	DURATION
BEGINNER	2	25 sec
INTERMEDIATE	3	25 sec
ADVANCED	3	30 sec

soleus

tibialis posterior

peroneus longus

peroneus brevis

flexor hallucis longus

Keep your heel on the ground.

CAUTION
Keep your upper body perpendicular to the ground while doing this exercise and hold onto a solid support.

INDICATION
For walkers, long-distance runners, swimmers, cyclists, soccer players, and other athletes who, due to the nature of their sports, work the muscles involved in plantar flexion of the ankle particularly intensely and for long periods of time.

Forced Dorsal Flexion

START

Stand at the edge of a step, stair, curb, or something similar that you have access to during your training. Advance one foot, place the heel on the edge of the step, and lower the front of the foot, placing your ankle in plantar flexion. Hold the front of your foot with both hands.

TECHNIQUE

Pull on the front of your foot without raising your heel from the support. This will produce maximum dorsal flexion of the ankle and thus stretch the soleus. Hold the tension for the time appropriate to your level and your goals.

peroneus longus

peroneus brevis

tibialis posterior

soleus

flexor hallucis longus

Keep your heel anchored on the original support point.

Starting Position

LEVEL	SETS	DURATION
BEGINNER	2	25 sec
INTERMEDIATE	3	25 sec
ADVANCED	3	30 sec

CAUTION

Plant your heel solidly on the surface, so it doesn't move from its anchor point as you perform this stretch.

INDICATION

For long-distance runners, walkers, and other athletes who, because of the nature of their sports, place particular demands on the muscles responsible for plantar flexion of the ankle.

Support on Stair

Starting Position

START
Stand in front of a step, stair, curb, or other object of a similar height, and rest the front of your foot on it. Keep your upper body perpendicular to the ground, your rear leg in line with your upper body, and your knee straight.

TECHNIQUE
Lean forward, keeping your upper body and your rear leg in a line, and perform maximum dorsal flexion of the forward ankle, without modifying the contact by the front of the foot. Your body weight will accentuate the dorsal flexion, producing an effective stretch.

tibialis posterior

flexor hallucis longus

peroneus longus

soleus

peroneus brevis

Touch the step with the front of your foot.

LEVEL	SETS	DURATION
BEGINNER	3	20 sec
INTERMEDIATE	5	30 sec
ADVANCED	6	40 sec

CAUTION
Use a solid step that will not move when you apply pressure.

INDICATION
For swimmers, cyclists, walkers, and long-distance runners, as well as for other athletes who experience excessive tension in the calf area.

Leg Shrugs

START
Kneel down and bend your knees completely, so that the rear of your thighs touches your calves. Lean your upper body forward slightly, place your arms beside your body, and rest your hands on the floor.

TECHNIQUE
Raise your knees from the floor while maintaining contact with your hands and the tips of your toes. Your body will move rearward a bit, and most of your body weight will be supported by your arms. Your ankles will achieve maximum plantar flexion, which will encourage stretching in the tibialis anterior.

tibialis anterior

Support most of your weight with your hands.

extensor digitorum longus communis (pedis)

extensor hallucis longus

Starting Position

LEVEL	SETS	DURATION
BEGINNER	2	20 sec
INTERMEDIATE	2	25 sec
ADVANCED	2	30 sec

CAUTION
Do this exercise on a mat or other cushioned surface whenever possible, so that you don't hurt your hands, knees, or ankles.

INDICATION
For long-distance and cross-country runners or for people who experience problems with the tibialis anterior, such as tendonitis, because of its intensive use in long, frequent workouts.

Seated Leg Crossover

Starting Position

tibialis anterior

extensor digitorum communis (pedis)

extensor hallucis longus

Pull on the front of your foot.

START
Sit down with one leg straight. Bend the opposite knee and cross that leg over the straight leg. Rest the ankle of the bent leg on your thigh and hold the front of your foot with one hand. The other hand can rest on the bent knee, and your back should be perpendicular to the floor.

TECHNIQUE
Pull on the front of your foot, accentuating the plantar flexion of your ankle without moving it from its support point. This will force the stretch of the tibialis anterior, which you should hold for the proper time for this exercise and your level.

LEVEL	SETS	DURATION
BEGINNER	2	20 sec
INTERMEDIATE	2	25 sec
ADVANCED	2	30 sec

CAUTION
Keep your ankle on the same location on your thigh throughout the exercise and keep it from rotating.

INDICATION
For long-distance runners, for athletes who train for long times or on uneven surfaces, and for people who experience tension in the front of the leg (the area between the knee and the ankle).

Prayer Position

START
Kneel down so that your knees and the tips of your toes touch the floor. Both lower limbs should be together and your hands should rest on your thighs. Keep your back erect and perpendicular to the floor.

TECHNIQUE
Hold one knee with one hand and pull on it, so that it comes off the floor. Your ankle should move from a neutral position to maximum plantar flexion, while you maintain the support on the tips of your toes. Lean your upper body forward slightly and use your hand on your thigh to help with stability.

Keep the tips of your toes in contact with the floor.

tibialis anterior

extensor digitorum communis (pedis)

extensor hallucis longus

peroneus anterior

LEVEL	SETS	DURATION
BEGINNER	2	20 sec
INTERMEDIATE	2	25 sec
ADVANCED	2	30 sec

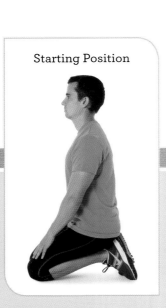

Starting Position

CAUTION
Do this movement slowly to keep your balance and to change the weight distribution on the various supports as you perform the exercise.

INDICATION
For runners in long-distance or cross-country sports and for athletes who experience tension in the front of the leg.

Stretch on All Fours

LEVEL	SETS	DURATION
BEGINNER	2	20 sec
INTERMEDIATE	2	25 sec
ADVANCED	2	30 sec

Rest your buttocks on your heels.

tibialis anterior

extensor digitorum communis (pedis)

peroneus anterior

extensor hallucis longus

START
Get down on all fours, on your hands and knees, by bending the latter at an angle slightly less than 90°. As a result, your body will be a little farther to the rear than normal, with your hips slightly lower. The tips of your feet will be in contact with the floor, but they will not bear any weight.

TECHNIQUE
Bend your knees all the way, lowering your hips and resting the backs of your thighs on your calves. Your hands will move rearward along with your upper body, and your body weight will end up on your ankles and the tips of your toes. This will produce maximum plantar flexion in your ankles and the stretch in your tibialis anterior.

Starting Position

CAUTION	INDICATION
Move your weight slowly, so that you don't endanger your knees.	For runners and walkers whose sports require lengthy training or competition.

Dance Step

START

Place one foot about one step ahead of the other. Your toes should face straight ahead. You can rest your hands on your waist, with your elbows out to the side. Keep your knees straight and your upper body perpendicular to the floor.

TECHNIQUE

Bend at the hips and move your body forward. Your rear foot will change from flat on the floor to tiptoe, and your ankle will be in plantar flexion with your instep facing the floor. Lowering your upper body will accentuate the plantar flexion of the ankle and facilitate the stretch.

Starting Position

tibialis anterior

peroneus anterior

extensor digitorum communis (pedis)

extensor hallucis longus

LEVEL	SETS	DURATION
BEGINNER	2	20 sec
INTERMEDIATE	2	25 sec
ADVANCED	2	30 sec

CAUTION

Avoid doing this exercise quickly or abruptly, so that you don't lose your balance.

INDICATION

For long-distance runners, walkers, and cross-country runners, because of the tension to which they subject the tibialis anterior, and for other athletes who experience tension in the front of the leg.

Keep the tips of your toes on the floor.

Raised Rear Support

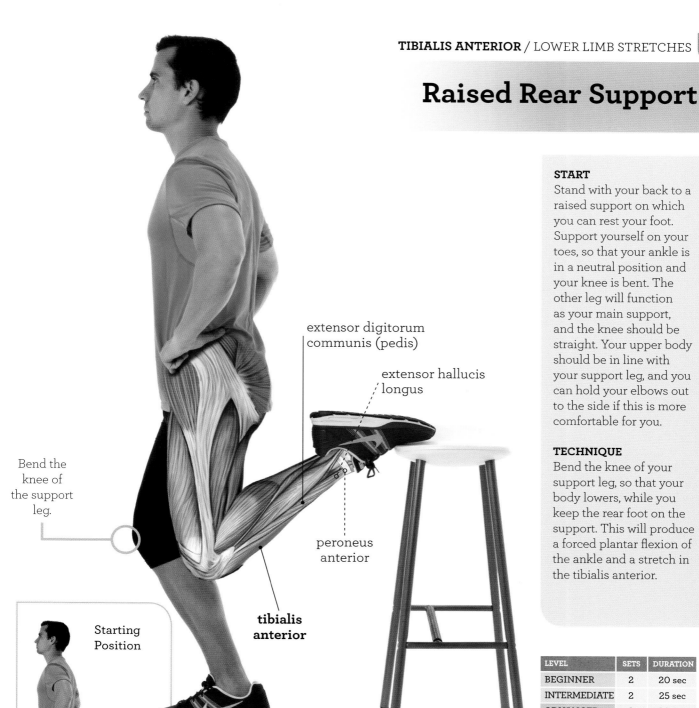

extensor digitorum communis (pedis)

extensor hallucis longus

Bend the knee of the support leg.

peroneus anterior

tibialis anterior

Starting Position

START
Stand with your back to a raised support on which you can rest your foot. Support yourself on your toes, so that your ankle is in a neutral position and your knee is bent. The other leg will function as your main support, and the knee should be straight. Your upper body should be in line with your support leg, and you can hold your elbows out to the side if this is more comfortable for you.

TECHNIQUE
Bend the knee of your support leg, so that your body lowers, while you keep the rear foot on the support. This will produce a forced plantar flexion of the ankle and a stretch in the tibialis anterior.

LEVEL	SETS	DURATION
BEGINNER	2	20 sec
INTERMEDIATE	2	25 sec
ADVANCED	2	30 sec

CAUTION
Choose a solid support and use your hands to steady yourself if necessary.

INDICATION
For long-distance runners and for athletes in sports where the roughness or unevenness of the terrain are significant factors.

Standing Leg Crossover

START

Cross one leg over the other, so that the rear leg supports most of your body weight and the corresponding knee is straight. The crossed leg is bent at the knee, so you can rest the tips of your toes on the floor next to the outside of the other foot.

TECHNIQUE

Bend the knee of the leg holding most of your weight, thereby forcing the other leg to bend farther, but without moving your feet from their original support points. The ankle of your crossed leg will have to straighten, and your sole will end up perpendicular to the floor.

tibialis anterior

extensor digitorum communis (pedis)

peroneus anterior

extensor hallucis longus

The crossed foot is on tiptoes.

Starting Position

LEVEL	SETS	DURATION
BEGINNER	2	15 sec
INTERMEDIATE	2	20 sec
ADVANCED	2	25 sec

CAUTION

Do the movement slowly, without moving your feet from their original contact points, and be sure to keep your balance as you do the exercise.

INDICATION

For long-distance and cross-country runners and for athletes in sports with uneven terrain. Also for walkers and for people involved in sports of medium or long duration.

Seated Ankle Inversion

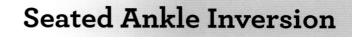

peroneus longus

peroneus brevis

peroneus anterior

extensor digitorum communis (pedis)

Force the ankle inversion.

START
Sit on a fairly high support such as a chair, a stool, a bench, or the like. Cross one leg over the other, so that the area above your ankle rests on the opposite thigh. Hold the outside of the foot with one hand and rest the other hand on your knee.

TECHNIQUE
Pull on the outside of your foot, causing ankle inversion, until you feel the tension from the stretch in the outside of your leg. The feeling of tension is not as prominent as in other muscle groups, such as the ischiotibials or the quadriceps femoris, but this does not mean that the stretch is not happening.

Starting Position

LEVEL	SETS	DURATION
BEGINNER	2	15 sec
INTERMEDIATE	2	20 sec
ADVANCED	2	25 sec

CAUTION
Do not exceed the recommended times for this exercise, because peroneus muscles are largely responsible for ankle stability.

INDICATION
Especially for cross-country and trail runners, but also for other running or walking sports that take place on uneven surfaces.

Inverted Support

START

Stand with your feet slightly farther apart than shoulder-width, with your toes pointing straight ahead. Keep your knees straight and your elbows out to the side for comfort.

TECHNIQUE

Move your body slightly to one side to produce ankle inversion, but without moving from any of your original contact points. Hold the tension in the outside of your leg for a few seconds.

Starting Position

peroneus longus

extensor digitorum communis (pedis)

peroneus brevis

LEVEL	SETS	DURATION
BEGINNER	2	15 sec
INTERMEDIATE	2	20 sec
ADVANCED	2	25 sec

CAUTION

Do the movement slowly and spread your weight between the two support points, so that you don't twist your ankle.

INDICATION

For runners who train or compete on uneven ground, such as trail runners.

Make contact with the outside of your foot.

Straight-leg Pull

START
In a sitting position, hold one leg out straight, with your other knee bent. Lean your upper body forward and hold the toe of the forward foot. You can place your other hand on your thigh.

TECHNIQUE
Use the hand holding the forward foot to produce ankle inversion. The rest of your body should stay in the same position, so that the movement will be very limited and the feeling of tension will not be pronounced, even though the stretch is happening.

Hold your foot by the toes.

peroneus longus

peroneus brevis

extensor digitorum communis (pedis)

Starting Position

LEVEL	SETS	DURATION
BEGINNER	2	15 sec
INTERMEDIATE	2	20 sec
ADVANCED	2	25 sec

CAUTION
Lean your upper body forward without exaggerating the bend in your spine, especially if you experience or have experienced spinal pain or discomfort.

INDICATION
For runners, walkers, and hikers who train or compete on irregular surfaces.

Forward Stride with Knee Bend

START

Place one foot ahead of the other, distributing your weight equally on both. Your knees should be straight and your upper body perpendicular to the floor. Your hands can hang next to your body or rest on your hips for greater comfort.

TECHNIQUE

Without moving your feet from their original support points, bend your knees to about 135°, moving your center of gravity forward and lower. The forward foot should remain flat on the floor, but the rear one should rest only on your toes to force the extension.

Starting Position

Touch the ground with your toes.

LEVEL	SETS	DURATION
BEGINNER	2	25 sec
INTERMEDIATE	3	30 sec
ADVANCED	3	35 sec

CAUTION

Concentrate on keeping your balance while doing this exercise.

INDICATION

For runners who experience or have experienced plantar fasciitis and pain in the sole of the foot. Also for competitive walkers, cross-country runners, and other long-distance runners.

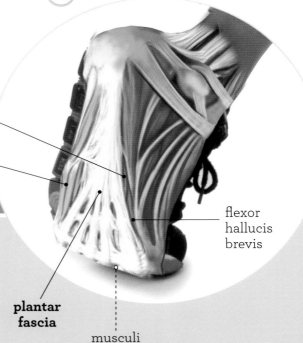

flexor digitorum brevis

flexor digiti minimi brevis pedis

flexor hallucis brevis

plantar fascia

musculi lumbricales pedis

Toe Support

Starting Position

Keep your heel raised and support yourself on your toes.

START
Place one foot behind the other at a short distance between the heel of one and the toes of the other. The forward leg should have a minimal degree of bend in the knee, and the rear leg will be nearly totally straight. Both feet should be flat on the floor.

TECHNIQUE
Bend both knees to about 90°, so that your center of gravity lowers, and raise the heel of the rear foot. This foot will shift to toe support and extension, which will cause the stretch in the plantar fascia.

flexor digitorum brevis (pedis)

flexor hallucis brevis

flexor digiti minimi brevis

musculi lumbricales pedis

plantar fascia

LEVEL	SETS	DURATION
BEGINNER	2	25 sec
INTERMEDIATE	3	30 sec
ADVANCED	3	35 sec

CAUTION
This stretch involves no risk other than losing your balance, so all you have to do is start with a solid position and do the exercise gradually.

INDICATION
For runners, walkers, and hikers, especially if they cover long distances and practice their sports on irregular terrain.

One-hand Pull

Keep your lower spine straight.

START

Stand with one foot a short distance ahead of the other. Bend your knees slightly and lean your upper body forward until it is nearly parallel to the floor. Hold the toes of the forward foot with one hand.

TECHNIQUE

Pull on the toe of your foot to produce extension in your toes and produce tension in the plantar fascia. Be sure to keep your lower spine straight, so that you don't injure your back while doing this exercise. If you feel discomfort, you can do the pull in a sitting position.

Starting Position

LEVEL	SETS	DURATION
BEGINNER	2	25 sec
INTERMEDIATE	3	30 sec
ADVANCED	3	35 sec

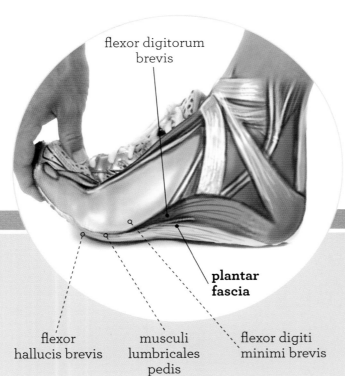

flexor digitorum brevis

plantar fascia

flexor hallucis brevis

musculi lumbricales pedis

flexor digiti minimi brevis

CAUTION

Do not apply excessive or sudden tension to the toes, because their joints are small and fragile.

INDICATION

For runners, walkers, and hikers whose activities go on for extended periods, especially if they frequent irregular terrain.

Support on Knees

Starting Position

Keep your toes extended.

flexor digitorum (pedis)

flexor digiti minimi brevis

flexor hallucis brevis

plantar fascia

musculi lumbricales pedis

START
Kneel down with your knees bent to about 80° and keep your upper body in line with your thighs. Your feet should rest on tiptoes, with the toes in a neutral position. You can rest your hands on your hips or let them hang at your sides.

TECHNIQUE
Increase the bend in your knees, so that the backs of your thighs touch and rest on your heels. Your body weight will move rearward. When the movement is over your feet, it will cause toe extension and a stretch in the plantar fascia.

LEVEL	SETS	DURATION
BEGINNER	2	25 sec
INTERMEDIATE	3	30 sec
ADVANCED	3	35 sec

CAUTION
If you feel pain in your toes, shorten the duration of the exercise or choose any of the other three exercises in this chapter intended for the plantar fascia.

INDICATION
For runners, walkers, and hikers, especially if they travel over irregular terrain, as well as for athletes who experience or have experienced plantar fasciitis, or pain in the front of the foot.

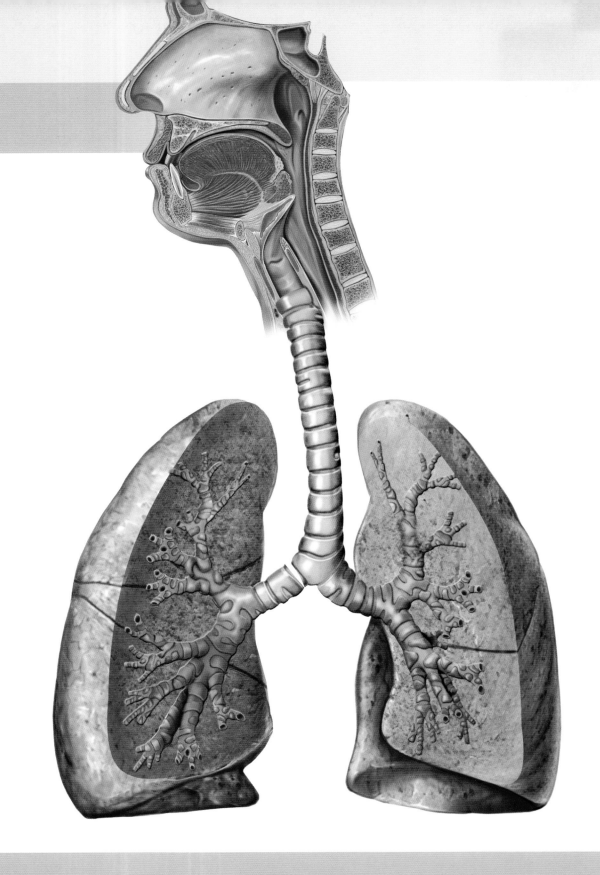

RESPIRATORY
MUSCLE
STRETCHES

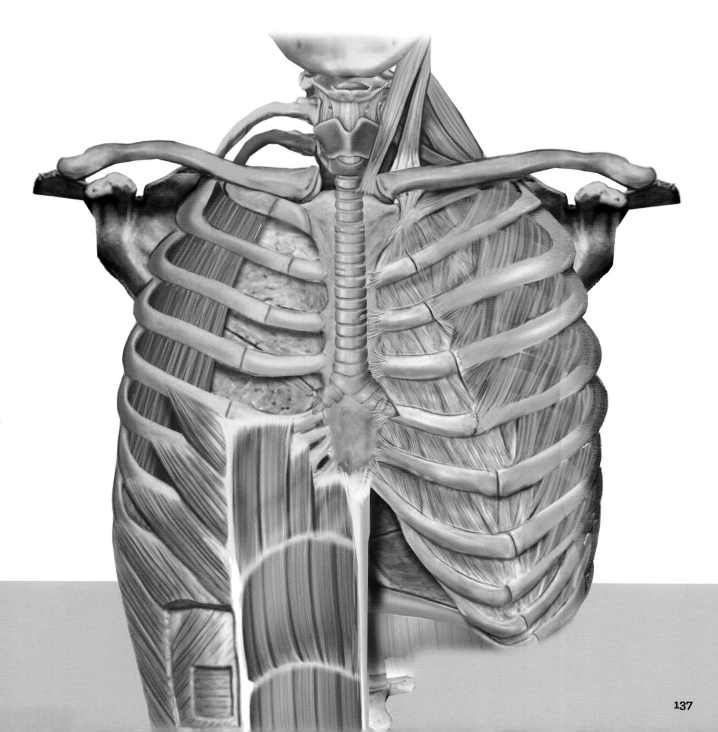

THE BASICS OF RESPIRATORY MUSCLE STRETCHES

Good conditioning of the muscles used in breathing largely determines the efficiency of the runner's respiratory mechanism and performance level. In endurance events, it is crucial to deliver sufficient oxygen to the muscles, so knowing how to breathe during exertion and doing it correctly will be a decisive factor.

Many respiratory exercises involve expansion of the rib cage.

Proper breathing is a crucial element for many runners, as well as for all athletes who play sports in which endurance is an important factor. A sprinter who runs the 100 meters or a shot-putter can become a champion in these sports, even if his or her breathing is inadequate. The reason is simple: in all short events, fatigue does not set in, so the oxygen supply is not a limiting factor. But, at the opposite end of the spectrum, if you are a marathoner, no matter how strong your muscles may be, you will not achieve good results if you do not manage to breathe efficiently in order to deliver to your muscles the oxygen and the energy that they need. Here is a very simple example (though slightly exaggerated): most of us, unless we suffer from some injury or disability, can run 100 meters, or even 200 meters, without breathing even once, and our time will not reflect that fact. Now try to run a marathon without breathing at all. If you can do that, you surely are the most efficient running machine that has ever existed! (Don't try this, of course.)

As we have already seen, in sports where the physical activity goes on for several minutes or more, it is essential to deliver to the working muscles a sufficient quantity of oxygen, so that they can continue performing. Otherwise, fatigue sets in, and the athlete has to stop.

Precisely the same situation occurs in running. In addition, most runners, except for the sprinters, keep at it for much longer than what we would consider a short time. Furthermore, endurance and superendurance events are increasingly well-attended: as a result, citizens' races, triathlon competitions, and other events are spreading and growing.

So, to participate in these events with some expectation of success, it is necessary to use good breathing tech-

The diaphragm stretch is one of the basic respiratory exercises.

niques, and the muscles involved in the respiratory function have to be at least as fit as your legs.

As you probably already know, the lungs do not fill themselves magically. Rather, they are in contact with the inside of the rib cage and, when this expands through the activation of certain muscles, so do the lungs. One proof of this is that, when a pneumothorax occurs, the lung loses contact with the rib cage. Then, no matter how much we expand the lung, it will remain retracted and it will not fill with air. Not only are there muscles that allow air to enter the lungs, but there are also muscles that force it out if we have to breathe intensely—for example, if we blow out a candle or blow up a balloon. The muscles that are responsible for breathing need to be in good condition if they are to be used intensely to get through a middle- or long-distance run, just as the pectorals must be if we hope to perform a heavy bench press. For this reason, it is useful to do breathing exercises, as well as regularly train and stretch these muscles, so that they function as completely as possible. The expansion and contraction of the thoracic cage must reach its maximum in every breathing cycle, so that we can use more oxygen. In addition, it is useful to know how to breathe, and there are several recommendations that we can make for people who have chosen running as an interest, hobby, or even passion.

■ Breathe in through your mouth and nose, and exhale in the same way, generally in two stages, and use a specific stride cadence as a reference. It's most usual to use a 2/2 rhythm. In other words, inhaling spans two strides and exhaling spans another two.

■ Use abdominal breathing, favoring the action of the diaphragm, which will allow you to take maximum advantage of your lung capacity and take in more oxygen with every breath.

This advice is useful, but all runners will find variations in timing and form that are best suited to their manner of running.

Finally, keep in mind that the stretches offered below are static ones. Thus, they should be done in accordance with the guidelines established in the preceding chapter.

Rib Cage Expansion

START

Place your hands on your torso, so that your palms are positioned near the last rib on each side. Keep your back perpendicular to the floor and breathe slowly, deeply, and calmly.

TECHNIQUE

Breathe in deeply and place your fingers, but not your thumbs, underneath your ribs. Let the air out and try to maintain the expansion of your thoracic cage by pulling with your fingers, so that your ribs do not sink back to their original position.

LEVEL	SETS	DURATION
BEGINNER	4	5 sec
INTERMEDIATE	5	8 sec
ADVANCED	5	10 sec

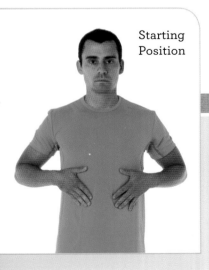

Starting Position

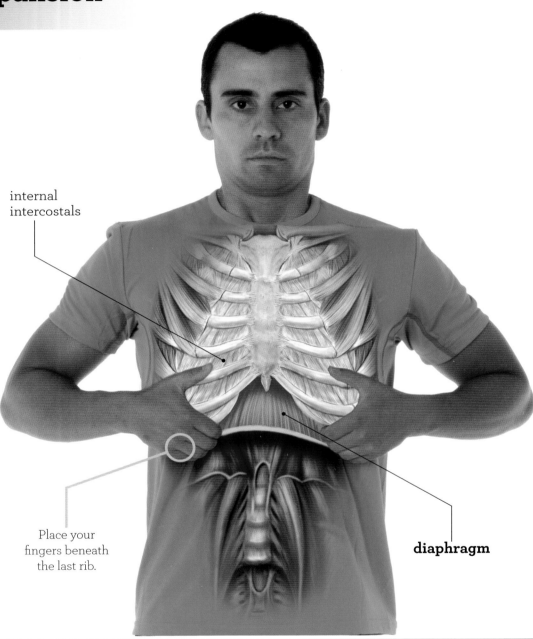

internal intercostals

Place your fingers beneath the last rib.

diaphragm

CAUTION

Sometimes, for various reasons, it may be difficult or painful to pull on the underside of your ribs with your fingers. If you experience this, reduce the duration and the intensity of each set.

INDICATION

For long-distance and ultra-long-distance runners, because of the special importance of breathing in their sports and the use of the diaphragm in breathing.

Assisted Elbow Pull

LEVEL	SETS	DURATION
BEGINNER	3	20 sec
INTERMEDIATE	3	25 sec
ADVANCED	3	30 sec

START

To do this exercise, you will need an assistant. Sit on the floor with your legs together and your knees bent to about 90°, with your feet resting on the floor in front of you. Place your hands on the back of your neck and keep your elbows out to the side. Your assistant should take a position behind you and hold your elbows, kneeling on one knee and facing to the side, so that his or her thigh supports your back.

TECHNIQUE

Your assistant should pull your elbows toward the rear while you keep your back resting solidly against the support thigh. This will produce an expansion of the thoracic cage and a stretch in your pectoral muscles. Keep your hands on the back of your neck and communicate constantly with your assistant.

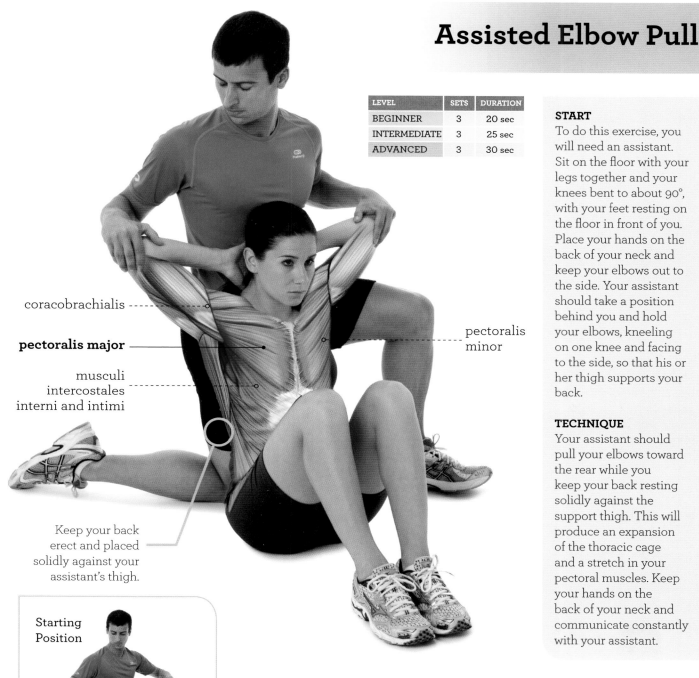

coracobrachialis

pectoralis major

musculi intercostales interni and intimi

pectoralis minor

Keep your back erect and placed solidly against your assistant's thigh.

Starting Position

CAUTION

Make sure that your assistant does the stretch slowly and is attentive to your instructions, knowing when to continue stretching and when to stop.

INDICATION

For runners and other athletes who take part in long-distance and ultra-long-distance events, such as swimmers, cyclists, walkers, and triathletes.

Swimmer's Position

START
Stand with your back erect and place your arms in front of you at a 90° angle with your upper body. Keep your elbows straight and place one hand on top of the other.

TECHNIQUE
Try to move the tips of your finger forward without changing the position of your body. To do this, you will have to move your shoulders forward, hollow your chest, and stretch the upper part of your back, where you will feel tension. You should hold the stretch for a few seconds.

Move your hands toward the front.

rhomboid

teres major

serratus anterior

LEVEL	SETS	DURATION
BEGINNER	3	20 sec
INTERMEDIATE	3	25 sec
ADVANCED	3	30 sec

Starting Position

CAUTION
This stretch involves no difficulty or risk of any kind, so you merely need to concentrate on moving your shoulders forward relative to your spine until you feel the tension in the top part of your back.

INDICATION
For long-distance and ultra-long-distance athletes, whether runners, cyclists, walkers, swimmers, triathletes, or devotees of other sports, because of the importance of proper breathing in these sports.

Leg Pull

Starting Position

LEVEL	SETS	DURATION
BEGINNER	3	20 sec
INTERMEDIATE	3	25 sec
ADVANCED	3	30 sec

START
Sit on something low, such as a step, or on the floor, if nothing else is available. Bend your knees to about 90°, keeping your legs and feet together. Lean your upper body forward, so that your chest touches your thighs, and cross your hands beneath your legs.

TECHNIQUE
Hold each elbow with the opposite hand beneath your thighs, which will limit your reach. Pull toward the rear with your body until your forearms encounter resistance from your thighs, your chest rounds inward, and your shoulders are ahead of the middle of your upper back.

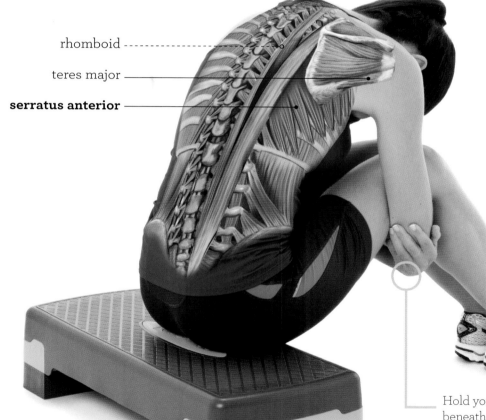

rhomboid

teres major

serratus anterior

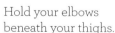

Hold your elbows beneath your thighs.

CAUTION
Even though this stretch involves no risk, there is a tiny possibility that you will experience some discomfort in your lower back. If so, replace this exercise with the Swimmer's Position (No. 97).

INDICATION
For long- or ultra-long-distance athletes, because of the importance of breathing in their sports. This group includes walkers, triathletes, cyclists, marathoners, long-distance swimmers, and cross-country runners.

Neck Lean

START

Starting from a sitting or standing position, place one hand on the top of your chest and collarbone. This hand should be planted firmly to serve as a stop to prevent the collarbone from rising. You can hold your forearm with the other hand in order to exert greater pressure.

TECHNIQUE

Lean your head to the side opposite the collarbone you are holding in place. Simultaneously, raise your chin slightly. When you perform this movement, also press down with your hand on your collarbone, trying as hard as you can to keep it from rising.

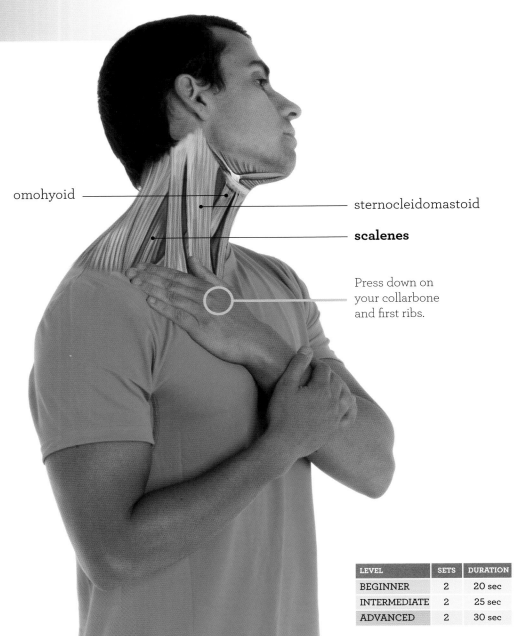

omohyoid

sternocleidomastoid

scalenes

Press down on your collarbone and first ribs.

Starting Position

LEVEL	SETS	DURATION
BEGINNER	2	20 sec
INTERMEDIATE	2	25 sec
ADVANCED	2	30 sec

CAUTION

Perform this movement slowly, so that you can feel the tension in the front and side of your neck. If you feel pain in your cervical vertebrae, or unsteadiness or dizziness, stop the stretch immediately.

INDICATION

For athletes who take part in long- and ultra-long distance training sessions or events, such as walkers, marathoners, triathletes, long-distance swimmers, and other endurance runners.

Chin Raise

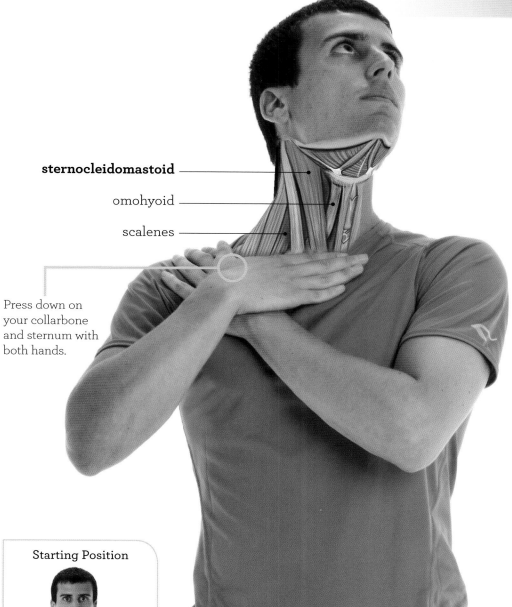

sternocleidomastoid

omohyoid

scalenes

Press down on your collarbone and sternum with both hands.

START
Cross both hands and place them on the third proximal of one collarbone and the top of your sternum. Apply pressure to this area, look straight ahead, and keep your back erect.

TECHNIQUE
Raise your chin, as if you were looking at the sky, as you slightly turn your neck toward the side opposite your hands, and press down with them on your collarbone and sternum. The tension in the front of your neck, which you will quickly feel in this exercise, indicates that you are doing the exercise correctly.

LEVEL	SETS	DURATION
BEGINNER	2	20 sec
INTERMEDIATE	2	25 sec
ADVANCED	2	30 sec

Starting Position

CAUTION
Stop doing this stretch if you feel any discomfort in the cervical area, dizziness, or nausea.

INDICATION
For resistance athletes whose sports make breathing a decisive factor, such as walkers, marathoners, triathletes, swimmers, divers, and cyclists.

Basic Warm-up Routine

This routine includes some of the stretches that you should do in the warm-up prior to any running event or intense training session. As you probably already know, the best stretches at this time are dynamic ones, which you should always do after a five- to ten-minute warm-up run at a very gentle pace.

7 page 36

HIP ROTATIONS

12 page 41

SLALOM

14 page 43

ASSISTED HIP FLEX

16 page 45

GOOSE STEP

13 page 42

ANKLE ROLLS

15 page 44

STAIR STEPS

This routine includes stretches to be done after athletic activity, whether competition or training. It is made up of static stretches that contribute to relaxation and recovery by the muscles that experience significant fatigue during running events. Remember, you should avoid extreme stretches after moderate or intense athletic practice.

37 page 76
STATIC BUTTERFLY

43 page 82
STANDING REAR CROSSOVER WITH SUPPORT

44 page 83
KNIGHT'S POSITION

47 page 86
CROSSED LEG PULL

49 page 88
KNEE PULL TO CHEST

56 page 97
FLAMINGO POSITION

65 page 106
SEATED UNLATERAL HIP FLEX

69 page 110
SOLID SUPPORT PUSH

Complete Routine

This is a routine intended for use in specific flexibility sessions that you can do when you finish a light workout or in isolated sessions specifically for improving the range of movement in various joints. In this case, you will have to do a prior warm-up of the areas you intend to stretch. This routine is composed of static exercises that address the main muscle groups involved in running.

18 page 53

ANTEPULSION WITH CROSSED HANDS

25 page 60

UPPER BODY ROTATION WITH POLE

29 page 66

BILATERAL RETROPULSION

32 page 69

ASSISTED REARWARD PULL

34 page 71

ASSISTED RETROPULSION

39 page 78

ALTERNATING HIP ABDUCTION

42 page 81

STANDING REAR CROSSOVER

45 page 84

LOW STRIDE

49 page 88

KNEE PULL TO CHEST

55 page 96

UNILATERAL ON SIDE

57 page 98

KNIGHT'S POSITION WITH PULL

60 page 101

HIP FLEX WITH SUPPORT

73 page 114

BILATERAL PULL WITH TOWEL

82 page 123

SEATED LEG CROSSOVER

88 page 129

SEATED ANKLE INVERSION

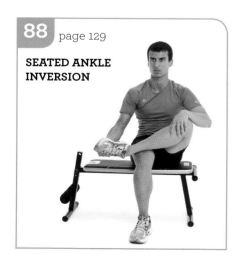

95 page 140

RIB CAGE EXPANSION

Alphabetical List of Muscles

Bibliography

Fields, Karl B., Burnworth, Craig M., and Delaney, Martha. *Should Athletes Stretch before Exercise?*, No. SSE#104, 2009, found at *https://ce.gssiweb.com/Article(Detail.aspx?articleID=736.*

Nelson, Russel T. and Bandy, William D. *Eccentric Training and Static Stretching Improve Hamstring Flexibility of High School Male*, No. 39(3), 2004. Found at *http://www.ncbi.nlm.nih. gov/pmc/articles/PMC422148/#_ffn_sectitle.*

Pereles, Daniel, Roth, Alan, and Thomson, Darby, J. S. *A Large, Randomized, Prospective Study of the Impact of a Pre-run Stretch on the Risk of Injury in Teenage and Older Runners.* CAQ Sports Medicine, 2009.

Shrier, Ian. "Does Stretching Improve Performance?: A Systematic and Critical Review of the Literature." *Cin J Sport Med*, Vol. 14, No. 5, 2004.

Vehm, David G. and Young, Warren B. "Effects of Running, Static Stretching, and Practice Jumps on Explosive Force Production and Jumping Performance." *The Journal of Sports Medicine and Physical Fitness*, Vol. 43, No. 1, 2003.